EXTRA LIFE

ALSO BY STEVEN JOHNSON

EXTRA
LIFE

A SHORT HISTORY OF
LIVING LONGER

Steven Johnson

RIVERHEAD BOOKS
NEW YORK
2021

RIVERHEAD BOOKS
An imprint of Penguin Random House LLC
penguinrandomhouse.com

Copyright © 2021 by Steven Johnson
Penguin supports copyright. Copyright fuels creativity, encourages
diverse voices, promotes free speech, and creates a vibrant culture. Thank you
for buying an authorized edition of this book and for complying with copyright
laws by not reproducing, scanning, or distributing any part of it in any
form without permission. You are supporting writers and allowing
Penguin to continue to publish books for every reader.

Riverhead and the R colophon are registered trademarks
of Penguin Random House LLC.

The PBS Logo is a registered trademark of the Public Broadcasting Service and
used with permission. All rights reserved.

Grateful acknowledgment is made for permission to reprint the following images:

Page xii: "Emergency hospital during influenza epidemic, Camp Funston, Kansas"
(NCP 001603). OHA 250: New Contributed Photographs Collection.
Otis Historical Archives, National Museum of Health and Medicine.

Page 4: Courtesy of the University of Toronto; http://hdl.handle.net/1807/34767.

Page 156: Photo of replica of Heatley's apparatus for the extraction and purification of
penicillin courtesy of the Science Museum / Science and Society Picture Library.

Library of Congress Cataloging-in-Publication Data

Names: Johnson, Steven, 1968– author.
Title: Extra life : a short history of living longer / Steven Johnson.
Description: New York : Riverhead Books, 2021. |
Includes bibliographical references and index.
Identifiers: LCCN 2020033229 | ISBN 9780525538851
(hardcover) | ISBN 9780525538875 (ebook)
Subjects: MESH: Public Health Practice—history | Life Expectancy—history |
Delivery of Health Care—history | Socioeconomic Factors—history |
Technology—history
Classification: LCC RA418 | NLM WA 11.1 | DDC 362.1—dc23
LC record available at https://lccn.loc.gov/2020033229

Printed in the United States of America
1st Printing

BOOK DESIGN BY MEIGHAN CAVANAUGH

For my mother

CONTENTS

INTRODUCTION
TWENTY THOUSAND DAYS

The stretch of the Kansas River just north of the town of Junction City has military roots that date back to 1853, when a post was established to protect travelers heading west in the years after the California gold rush. Within a few decades, the site became known as Fort Riley, and for a time it hosted the United States Cavalry School. In 1917, as the armed forces began preparing for the American entry into World War I, a small city with a population of fifty thousand was erected practically overnight to train Midwestern soldiers headed overseas to fight in the Great War.

Camp Funston, as it was called, contained around three thousand temporary buildings—the predictable barracks and mess halls and commanders' offices, but also general stores, theaters, and even a coffeehouse. The pop-up city had meaningful amenities for the young recruits—one soldier wrote home with an account of hearing a symphony entertaining the troops at Camp Funston—but the temporary nature of the structures meant that most of the quarters were barely insulated. The first winter of the camp's existence was an unusually frigid one, forcing the soldiers already bunked together in tight quarters to cluster even closer around the stoves in the barracks and the mess halls.

As winter drew to a close in early March 1918, a twenty-seven-year-old private named Albert Gitchell presented himself at the infirmary complaining of muscle pain and fever.[1] Gitchell was a butcher by trade, and had been serving as a mess cook at Camp Funston, preparing food for hundreds of his fellow soldiers in training. Doctors diagnosed his illness as influenza and dispatched him to the contagion ward in hopes of preventing the spread of the disease, but the intervention proved to be too late. Within a week,

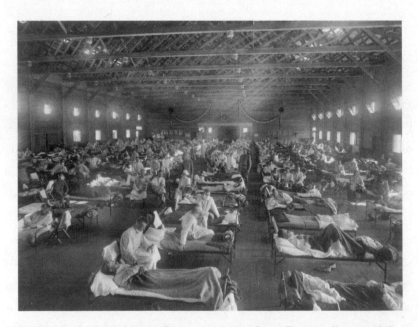

Emergency hospital during influenza epidemic, Camp Funston, Kansas, 1918

hundreds of other residents of Camp Funston reported influenza symptoms. By April more than a thousand soldiers at Camp Funston had been hospitalized. Thirty-eight of them died, a surprisingly high number for a disease that usually threatened only the very young and the very old.

The crowded infirmaries and the bodies piled in the morgue at Camp Funston were early clues that something unusual was happening at that Kansas army base. But what was really happening there would not be visible to scientists until the development of electron microscopy decades later. Inside the lungs of Albert Gitchell, a sphere covered in spikes fastened itself to a cell lining on the surface of the young private's respiratory tract. The sphere burrowed through the cell membrane into the cytoplasm, fused its own limited strand of genetic code with Gitchell's, and began making copies of itself. Within about ten hours, the cell was teeming with newly replicated spheres, stretching the membrane to a breaking point, until, in one catastrophic instant, the cell exploded, releasing hundreds of thousands of new spheres into the mess cook's respiratory tract. Some of those spheres were coughed or sneezed into the air of the mess hall and the barracks. Others remained in the lungs of private Gitchell, latching onto other cells with the same brutal machinery of self-replication.

The doctors at Camp Funston had no way of knowing it at the time, but those spheres attacking Albert Gitchell's lungs constituted a new strain of the H1N1 virus that would come to terrorize the entire world over the next two years in the pandemic commonly referred to as the Spanish flu. Just as the virus itself replicated in the respiratory tracts of the soldiers, the scene at Camp Funston would be replicated at military bases around the globe in the coming months, fueled by the steady flow of soldiers across the

United States and onto the European front lines. American troops brought the virus to the military port at Brest, on the northwest edge of Brittany in France; the virus then erupted in Paris in late April. Italy soon followed. On May 22, the Madrid newspaper *El Sol* reported that "a sickness which has not yet been diagnosed by doctors"[2] was ravaging the garrisons of Madrid. By the end of May the virus was in Bombay, Shanghai, and New Zealand.

The strain of H1N1 that encircled the globe in the spring of 1918 spread at an alarming rate compared to most influenzas; it passed readily from person to person, and successfully set off chains of cell rupture in the lungs of many of those people. But it wasn't particularly lethal. The flu's ability to race around the world in such a short amount of time—all those self-replicating spheres in all those lungs—was formidable. And yet many of those lungs recovered from the attack. In the technical language, the strain displayed a high *morbidity* rate combined with a more modest rate of *mortality*. It made copies of itself with fearsome skill, but it tended to let its hosts survive the encounter.

The strain of H1N1 that erupted in the fall of 1918 would not be so generous.

To this day, scientists debate why the second wave of the Spanish flu in 1918 proved to be so much more virulent than the virus that first emerged in the spring. Some argue that the two waves were propelled by different variations of H1N1; others believe that the two different strains encountered each other in Europe and somehow combined into a new, more lethal variant. Others believe that the initial wave was weaker because the virus had only recently jumped from animal hosts to human ones and required a number of months to properly adapt to its new habitat in the respiratory tracts of *Homo sapiens*.[3]

Whatever the underlying cause, the wake of death left behind in the second wave was staggering. In the United States, the new threat first became apparent at Camp Devens, an overcrowded military base on the outskirts of Boston. By the third week of September, one fifth of the camp's population had contracted influenza, a morbidity rate that exceeded that of the H1N1 outbreak at Camp Funston. But it was the mortality rate that shocked the medical staff at Camp Devens. "It is only a matter of a few hours then until death comes," one of the army physicians wrote:

> It is horrible. One can stand it to see one, two or twenty men die, but to see these poor devils dropping like flies. . . . We have been averaging about 100 deaths per day. . . . Pneumonia means in about all cases death. . . . We have lost an outrageous number of Nurses and Drs., and the little town of Ayer is a sight. It takes special trains to carry away the dead. For several days there were no coffins and the bodies piled up something fierce. . . . It beats any sight they ever had in France after a battle. An extra long barracks has been vacated for the use of the Morgue, and it would make any man sit up and take notice to walk down the long lines of dead soldiers all dressed and laid out in double rows.[4]

The devastation at Camp Devens would soon be followed by even more catastrophic outbreaks around the world. In the United States, almost 50 percent of all deaths would be attributable to influenza over the next year. Millions died on the front lines and in military hospitals in Europe. The mortality rate for those infected in parts of India approached 20 percent, an order of magnitude more lethal than the first-wave virus had been. Today the best estimates suggest that as many as a hundred million people died of influenza during the outbreak around the world. John Barry, author of the canonical account of the outbreak, *The Great Influenza*, puts that number in context: "Given the world's population in 1918 of approximately 1.8 billion, the upper estimate would mean that in two years—and with most of the deaths coming in a horrendous twelve weeks in the fall of 1918—in excess of 5 percent of the people in the world died."[5]

Mortality reports revealed another disturbing element of the pandemic: the H1N1 outbreak of 1918-19 was unusually lethal among young adults, normally the most resilient cohort in ordinary flu seasons. In the United States, as Barry notes, "the single greatest number of deaths occurred in men and women aged twenty-five to twenty-nine, the second greatest number in those aged thirty to thirty-four, the third greatest in those aged twenty to twenty-four. And more people died in each of those five-year groups than the total deaths among all those over age sixty."[6] In part, this unusual pattern was attributable to the virus's spread through the close quarters of military barracks and hospitals. Scientists also believe that a similar virus that emerged in 1900 had left a significant portion of the older population immune to the Spanish flu variant.

The unusual demography of the Spanish flu was clearly visible in the life expectancy charts that were ultimately calculated for the period. Everyone under the age of fifty saw a precipitous drop in expected life during the H1N1 outbreak, while the life expectancies of seventy-year-olds were unaffected by the pandemic. But overall, the story was bleak beyond imagination. In the United States, average life expectancy plunged an entire decade, practically overnight. India may have experienced the lowest known life expectancy of any human society in history, whether industrial or agricultural or hunter-gatherer. In England and Wales, where life expectancy had been rising for a half century, a virus amplified by war had undone it all in just three years. On the eve of World War I, life expectancy at birth—for the entire population, not just the elites—had risen to fifty-five. By the end of the twin catastrophes of world war and pandemic, a child born in England and Wales had a life expectancy of just forty-one years, not far from what those populations would have experienced during the Elizabethan era.

Even before those numbers were computed, as the H1N1 virus was still exploding cells in the lungs of human beings all around the world, the army scientist Victor Vaughan was analyzing the rough casualty counts coming in from the European front. "If the epidemic continues its mathematical rate of acceleration," he speculated in a handwritten letter, "civilization could easily disappear . . . from the face of the earth within a matter of a few more weeks."[7]

IMAGINE YOU ARE THERE at Camp Devens in late 1918, surveying the bodies stacked in the makeshift morgue. Or you are roaming the streets of Bombay, where more than 5 percent of the population has died of influenza over the past few months. Imagine touring the military hospitals of Europe, seeing the bodies of so many young men mutilated by the new technologies of warfare—machine guns and tanks and aerial bombers—and the cytokine storms of H1N1. Imagine knowing the toll this carnage would take on global life expectancy, with the entire planet lurching backward to health outcomes that belonged to the seventeenth century, not the twentieth. What forecast would you have made for the next hundred years, standing there at the end of the war and the pandemic, with the bodies piled around you? Was the progress of the past half century merely a fluke, easily overturned by military violence and the increased risk of pandemics in an age of global connection? Or was the Spanish flu the preview of an even darker outcome, as Victor Vaughn feared, where some rogue virus with an even more virulent "mathematical rate of acceleration" causes a global collapse of civilization itself?

Both grim scenarios seemed within the bounds of possibility, as the world slowly recovered from the double firestorm of the Great War and H1N1. And yet, amazingly, neither of those scenarios came to pass. Instead of tracking those bleak forecasts, what followed was a century of unexpected life.

The period from 1916 to 1920 marked the last point in which a major reversal in global life expectancy would be recorded. (During World War II, life expectancy did briefly decline but with nowhere near the severity of the collapse during the "Great Influenza.") The descendants of those English babies born in 1920 who could expect to live forty-one years now enjoy life expectancies in the eighties. And while Western societies captured most of the progress during the first half of this period, over the past few decades the developing world—led by China and India—has seen life expectancy rise faster than any society in history. Just a hundred years ago, the residents of Bombay or Delhi would be beating the odds simply by surviving into their late twenties. Today the average life expectancy in the Indian subcontinent is more than seventy years. Vaughn was right that there was an extraordinary "mathematical rate of acceleration" in our future. It just turned out to be a positive acceleration: more and more lives saved, not destroyed.

This march of progress is not unstoppable. The COVID-19 pandemic, which emerged almost precisely on the centennial anniversary of the end of the Great Influenza, has been a terrifying reminder that our globally interconnected world is more vulnerable than ever to fast-moving infections. To date, the COVID pandemic has reduced US life expectancy by about a year, and twice that in African-American communities. But for all its terror and tragedy, the 2020 pandemic also showcases the advances we have made over the century that has passed since 1918. The death toll from COVID-19 is, so far, less than 1 percent of that of the 1918 pandemic, on a planet with four times as many people. Some estimates suggest that more than a million lives were saved by the public interventions in the first half of 2020, despite the many early mistakes made during that period. But another virus might combine SARS-CoV-2's stealthy asymptomatic transmission with the much higher fatality rates of the 1918 virus, killing children and young adults as ruthlessly as the coronavirus kills the elderly. If we are going to avoid a health crisis on that scale, if we are going to continue the tremendous progress in extending human life, we need to understand the forces that drove such momentous change over the past hundred years—not just to celebrate those achievements, but also to build on them.

THE MACRO STORY of human health in the century that has passed since the end of the Great Influenza can be told in three charts. Let's begin with the simplest one, tracking life expectancy in England back to the middle of the seventeenth century:[8]

LIFE EXPECTANCY

Shown is a period life expectancy at birth, the average number of years a newborn would live if the pattern of mortality in the given year were to stay the same throughout its life.

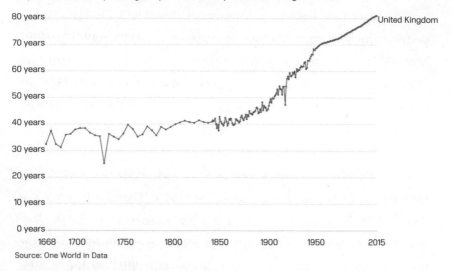

Source: One World in Data

British Life Expectancy at Birth, 1668–2015

There may be no more important chart that captures what has happened to the human race—and to the planet itself—than this one. In the early 1660s, when people first started tinkering with the idea of calculating life expectancies, the average British person lived just over thirty years. A child born in the United Kingdom today can expect to live a full fifty years longer than that. And that extraordinary upward slope has been repeated again and again around the world. All the advances of the last three or four centuries—the scientific method, the medical breakthroughs, the public health institutions, the rising standards of living—have given us about twenty thousand extra days of life on average. And billions of people who never would have lived to see adulthood or have their own children were endowed with those most precious of gifts.

There are few measures of human progress more astonishing

than this. From a long-term perspective, those extra twenty thousand days should be running as a headline in every newspaper, every day. But, of course, that extra life span almost never appears on the front page of newspapers, because it is a story almost entirely free of the traditional dramatic elements that drive the news cycle. It is the story of progress in its usual form: brilliant ideas and collaborations unfolding far from the spotlight of public attention, setting in motion incremental improvements that take decades to show their true magnitude. And so the news understandably chooses to focus on the short-term fluctuations: an upcoming election, a celebrity scandal, all the surface tremors that distract us from the movement of the underlying plates. Without that long view, we forget all the threats that terrorized our great-grandparents but were transformed into nonevents or manageable conditions so mundane that most of us never think about them at all. But that selective memory, as impressive as it is as a mark of progress, has an unfortunate side effect. By not thinking about those vanquished threats, we can be easily distracted from the underlying arc of progress—in basic human standards of health and social well-being—that has been the story of the past hundred years. And by not thinking about that past, we can't learn from it; we can't use that history to think more clearly about what advances to pursue in our current quest to extend the human life span; we can't use that history to prepare us for the unintended consequences those advances will inevitably bring; and we're less likely to trust the resources and institutions that we possess now to combat emerging threats like the COVID-19 pandemic. We have absurd conspiracy theories about Bill Gates planting microchips via mass vaccination or outright hostility directed at simple acts like mask-wearing in part because we have forgotten, as a culture, how much science and medicine and public health have improved the quality (and the length) of the average human life over the past few generations.

In a sense, human beings have been increasingly protected by an invisible shield, one that has been built, piece by piece, over the last few centuries, keeping us ever safer and further from death. It protects us through countless interventions, big and small: the chlorine in our drinking water, the ring vaccinations that rid the world of smallpox, the data centers mapping new outbreaks all around the planet. Those innovations and institutions rarely get the attention we regularly dole out to Silicon Valley billionaires or Hollywood stars—or even our military commanders. But the public health shield they have erected around us—measured, most clearly, in the doubling of life expectancy—is truly one of the greatest achievements in the history of our species. A crisis like the global pandemic of 2020–21 gives us a new perspective on all that progress. Pandemics have an interesting tendency to make that invisible shield suddenly, briefly, visible. For once, we're reminded of how dependent everyday life is on medical science, hospitals, public health authorities, drug supply chains, and more. And an event like the COVID-19 crisis does something else as well: it helps us perceive the holes in that shield, the vulnerabilities, the places where we need new scientific breakthroughs, new systems, new ways of protecting ourselves from emergent threats.

Most history books take as their central focus a person, or an event, or a place: a great leader, a military conflict, a city, or a nation. This book, by contrast, tells the story of a number: the rising life expectancy of the world's population, giving us an entire extra life in just one century. It is an attempt to understand where that progress came from, the breakthroughs and collaborations and institutions that had to come into being to make it possible. And it tries to answer that question rigorously: how many of those extra twenty thousand days came from vaccines, or randomized, controlled double-blind experiments, or the decrease in famines? The first mortality reports that enabled people to even think about life expectancy were designed to understand what was killing the people of seventeenth-century England. This book turns that inquiry on its head, and asks: What are the forces that now keep us alive?

AS IMPORTANT AS THE charts of overall life expectancy are, they do tell a slightly misleading story, one that often prompts fantasies of imminent immortality. When you look at the story of human life extension as an average—as a mean—it conveys a picture of runaway growth. Press fast forward and imagine how that trend line plays out over the coming century. If the same upward trend continued, the "average" person would live to 160.

But look at the story as a *distribution*, and the picture changes. The most significant reductions in mortality have happened in the first decade of life. Adults are certainly living longer than they did in the heyday of the Industrial Revolution—there are four times as many centenarians on the planet as there were in 1990—but the difference is not as dramatic as you might expect looking at the average life expectancy charts. Many people lived into their sixties and beyond two centuries ago. (Just think of the American founding fathers: Jefferson died at the age of seventy-three, while Madison, Adams, and Franklin all survived until their mid-eighties.) But the mortality rates of infants and young children have dropped precipitously. When you have a significant portion of the population dying at the age of five months or five years, those deaths pull the overall average life span down dramatically. But if most of those children survive into adulthood, average life expectancy spikes upward.

You can see the effect clearly by imagining a much smaller population of just ten people. If three of them die at the age of two—which is what you would expect in a society with 30 percent childhood mortality—but the rest live to seventy, then the mean life expectancy for the group is forty-nine years. Keep those three children alive and allow them to live to seventy like the others, and the overall average jumps twenty-one years to seventy. But in this scenario, the adults aren't living a day longer. It's just that the kids have stopped dying.

The oversized impact of early death is why demographers differentiate between the category of life expectancy "at birth" and life expectancy at other ages. In many societies, life expectancy at birth is significantly lower than life expectancy at fifteen or twenty, because the risks of death during infancy or early childhood are so severe. A newborn might only have a life expectancy of thirty, say, while a young adult could reasonably expect to live to fifty or beyond. In most modern societies, where childhood mortality is low, each year you survive detracts from the total subsequent years you can expect to live—one year older, one year closer to the end of your life. But in societies with high childhood mortality rates, the pattern is reversed: expected death gets further away as you age, at least through early adulthood.

All of which means that the iconic chart of runaway life expectancy growth should always be accompanied by a second chart that tracks the equally miraculous trends in childhood mortality:[9]

GLOBAL CHILD MORTALITY

Share of the world population dying and surviving the first 5 years of life

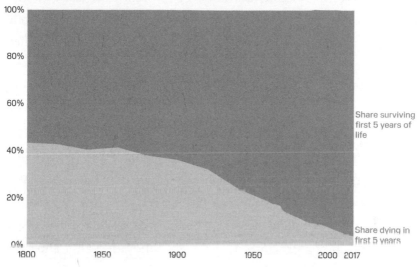

Global child mortality, 1800–2017

This book begins with these two simple but astonishing facts: as a species we have doubled our life expectancy in just one century, and we have reduced the odds of that most devastating of human experiences—the death of a child—by more than a factor of ten.

THIS BOOK IS, ultimately, a study in how meaningful change happens in society. A hundred years ago, as the body count from the Spanish flu was being tabulated, the notion that global life expectancy might reach into the seventies seemed almost preposterous. Today it is a reality. What changed between now and then? This is an old question, as it turns out. Almost as soon as demographers began noticing that life expectancy was increasing, scholars and public health experts began debating what was driving the change. Their varied investigations constitute one of the central threads of this book, because understanding the roots of positive change often turns out to be just as important as the specific breakthroughs that cause that change in the first place—in part because it allows you to discard false hypotheses or sham cures, and in part because it allows you to expand on the truly successful interventions, to bring their advances to a wider community.

A history book organized around a demographic trend—and not, say, the life of a famous leader, or a legendary battle at sea—poses some interesting organizational challenges. How do you tell a story with a thousand heroes? The chronological account gives you too much of a straight time line, one innovation after another: X-rays, antibiotics, polio vaccines. This book takes a different approach. It begins with an initial filter: defining the most meaningful *categories* of change that can explain the doubling of life expectancy over the past century. Some of those categories are obvious ones, beginning with the holy grail of the COVID era: vaccines. But defining some of the other categories is less clear-cut. What kind of metrics do you use? Perhaps there is some utilitarian ideal out there that we will someday be able to compute: years of expected life saved by a given idea. That would be the perfect data point to build the categories around. But that kind of calculation is hard to work out in the real world. To begin with, the exercise is by definition a counterfactual one. You're tracking lives saved, not deaths. Thanks to the invention of mortality reports and public health records, it is now quite easy to calculate how many people have been killed by a specific threat: pneumonia, say, or car accidents. In many parts of the world, that data is just a few clicks away, downloadable in an Excel file. But once you enter the hypothetical realm of alternate time lines—how many *would have been killed* had a specific intervention *not* been in place?—you enter a murkier terrain. One approach is to simply extrapolate from mortality rates before the intervention was widely utilized. For example, before the invention and widespread of adoption of seat belts, six people died for every 100,000 miles driven in the United States. If mortality rates had stayed at that level, then an additional ten million Americans would have died over the half century that has passed since then. But as we will see, the seat belt was one of a number of different factors that improved automobile safety during that period; air bags, Mothers Against Drunk Driving, crumple zones, and a thousand other small tweaks to car design and road safety also contributed.

The unavoidable fact about the history of human health is that the innovations that have driven progress are themselves almost always enmeshed in symbiotic relationships with other innovations. For instance, some attempts to gauge the life-saving effect of inventions throughout history have suggested that the humble toilet is responsible for saving more than a billion lives since its mass adoption began in the 1860s. There is, admittedly, something both plausible and instructive about this argument. The decline of waterborne diseases was one of the key driving forces in the first jump in life expectancy that appeared in Western industrialized countries, shortly after the toilet made its way into middle-class homes. And celebrating the life-saving virtues of the toilet reminds us that progress is often to be found in grittier inventions, not just the consumer tech we most frequently associate with so-called disruptive innovation. But for a toilet to actually improve health outcomes, it has to be connected to a functional sewer system that separates waste and drinking water. And for those expensive sewer systems to be built, we needed to replace the miasma theory of disease with an understanding of waterborne transmission. And for that to happen, we needed public health data and epidemiology to emerge as mature sciences. Yes, it is likely true that this entire complex—the physical object of the toilet, the public infrastructure of the sewers, the conceptual breakthroughs of the waterborne theory and epidemiology—saved more than a billion lives. But all the credit cannot go to the toilet alone.

Despite those real challenges, making rough estimates of the life-extending impact of recent interventions is still an exercise worth pursuing, because it helps us see what has worked in the past, and suggests road maps for what might work as future interventions. The haziness of the exercise means that it is best organized around orders of magnitude: innovations that saved millions of lives; those that saved hundreds of millions of lives; and the true giants of our extended life: the breakthroughs that saved billions. Organized this way, the story of humanity's extended life over the past few centuries looks something like this:

MILLIONS:

AIDS cocktail

Anesthesia

Angioplasty

Antimalarial drugs

CPR

Insulin

Kidney dialysis

Oral rehydration therapy

Pacemakers

Radiology

Refrigeration

Seat belts

HUNDREDS OF MILLIONS:

Antibiotics

Bifurcated needles

Blood transfusions

Chlorination

Pasteurization

BILLIONS:

Artificial fertilizer

Toilets/Sewers

Vaccines

Ranking the objects that saved the most lives—from the toilet to the bifurcated needle—has an undeniable tangible appeal as an exercise, and we will explore the stories behind many of these extraordinary breakthroughs over the coming pages. But there is also something misleading about viewing this history as a progression of *things*, each improving human health in new ways. Many of the changes that really matter cannot be reduced to a single object. Sometimes the crucial breakthroughs are meta-innovations: new ideas that make it easier to have new ideas, or to spread them. Sometimes these involve methods of manipulating information, or platforms that enable new forms of collaboration. Sometimes the meta-innovation is a new kind of institution, capable of amplifying life-saving ideas in a way that had been previously unimaginable. Sometimes the breakthrough is a conceptual advance in an unrelated field that expands the possibility space of health indirectly. These kinds of developments are more ephemeral than the classic eureka stories that make up most accounts of human progress, which is why we tend to be more familiar with stories about the accidental discovery of penicillin than developments like the creation of the Food and Drug Administration, which helped us separate genuine medicines from snake oil cures. But as we will see, the latter have had an enormous impact on human health, and often involve stories of quiet heroism and genius every bit as compelling as the traditional narratives of rogue investigators and their eureka moments.

In the end, I have organized this story of our extra life into eight main categories. The first is the concept of life expectancy itself, which turned out to be one of those innovations in the science of measurement that fundamentally changes the thing it is measuring. The others are: vaccines; data and epidemiology; pasteurization and chlorination; regulations and testing; antibiotics; safety technology and regulations; and antifamine interventions. Each category appears here as a chapter, telling the stories of the main agents who brought these new ideas into the world, and the people who fought to ensure that the ideas were adopted. Though I have tried to organize these chapters based on empirical public health data suggesting the innovations that had the biggest impact, the underlying categories are inevitably somewhat subjective ones. On occasion, I have erred on the side of less-familiar stories in the canon of human health, which means that a few more celebrated breakthroughs are only dealt with here in passing: Semmelweiss and the germ theory in the nineteenth century; the fight against AIDS in more recent years. But I have also tried to compile a representative sample that does justice to the overall trends.

Seen as a whole, the categories should convey a sense of the magnitude of the change itself—those twenty thousand extra days of life—and the vast range of talent, expertise, and collaboration that made them possible.

FOR ALL ITS EMPHASIS on progress and positive change, this book should not be mistaken for a victory lap, an excuse for resting on our laurels. It is by no means inevitable that the runaway growth of twentieth-century life expectancy will continue its upward march forever. As I write, the infection count of the COVID-19 pandemic is still growing; even before the outbreak, the United States had experienced an epidemic of opioid overdoses and suicides—the so-called deaths of despair—which had reduced life expectancies for the country for three years straight, the longest period of decline since the end of the Spanish flu.[10] Significant health gaps still exist between different socioeconomic groups and nations around the world. And ironically, the epic triumph of doubling life expectancy has created its own, equally epic set of problems for the planet. Consider the chart of global population since the agricultural revolution on the next page.[11]

It is no accident how closely the charts mirror the long view of life expectancy: millennia pass with almost no meaningful change, followed by a sudden, unprecedented spike over the past two centuries. The charts mirror each other because they are effectively

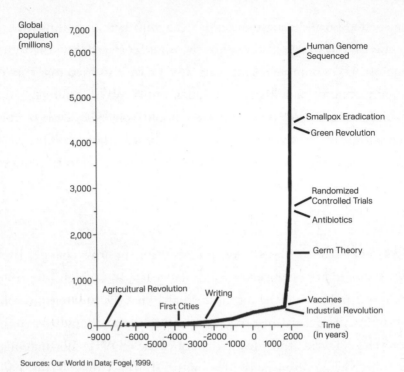

Sources: Our World in Data; Fogel, 1999.

Global Population Growth Since the Agricultural Revolution

mapping the same phenomenon. Demagogues sometimes rant about irresponsible birth rates in developing world countries, but the truth is the spike in global population is not caused by some worldwide surge in fertility. In fact, people are having fewer babies per capita than ever. What changed over the past two centuries, first in the industrialized world, then globally, is that people stopped dying—particularly young people. And by not dying, most of them lived long enough to have their own children, who repeated the cycle with their offspring. Increase the portion of the population that survives to childbearing years, and you'll have more children, even if each individual has fewer offspring on average. Repeat that pattern all over the world for six or seven generations and global population can grow from one billion to seven billion, despite declining fertility rates.

On one level, it is hard not to consider this to be fantastic news: all those children who would have died in infancy now allowed to have their own children, or to enjoy full lives well into adulthood. But it is equally difficult not to see something ominous in that runaway growth on the far right of the graph. That is not a shape that one normally sees in healthy natural systems with stable equilibria. That shape is the exponential march of cancer cells, or the spheres of the H1N1 virus self-replicating inside the respiratory tract. All those brilliant solutions we engineered to stop the growth of threats like H1N1 created a new, higher-level threat: ourselves. Many of the key problems we now face as a species are second-order effects of reduced mortality. For understandable reasons, climate change is usually understood as a second-order effect of the Industrial Revolution, but had we somehow managed to adopt a lifestyle powered by fossil fuels *without* reducing mortality rates—in other words, if we'd invented steam engines and coal-powered electrical grids and automobiles but kept global population at 1800 levels—climate change wouldn't be an issue at all. There simply wouldn't be enough humans to make a meaningful impact on carbon levels in the atmosphere.

And so the story of that simple number—life expectancy at birth—should not be understood as a story of unambiguous triumph. No change that momentous is ever purely positive in its effects. The doubling of life expectancy should, however, be understood as the most important development in human society over the last hundred years, in part because its effects are so intimate and so global at the same time. In just a few centuries, we managed to give ourselves an additional twenty thousand days of life. Billions of children who would have died in their first few years were able to grow into adulthood, to have children of their own. This book is the story of how that happened.

EXTRA LIFE

1

THE LONG CEILING

MEASURING LIFE EXPECTANCY

In the spring of 1967, a sociology graduate student from Harvard named Nancy Howell took a flight from Boston to Rome with her new husband, an anthropologist named Richard Lee. After a few days in Italy, they flew to Nairobi, where they met an academic friend of Richard's and visited the Hadza tribes living in the region. From there they flew to Johannesburg, where they loaded up on supplies and socialized with a few more researchers in the area.[1] They purchased a truck and drove north to the newly independent country of Botswana, picking up supplies in its new capital, then traveling northwest toward the swampy oasis of the Okavango Delta, recently flooded by seasonal rains. They rented a postbox in the town of Maun, the last outpost that would contain modern amenities like convenience stores and petrol stations. From Maun, they drove about 150 miles west, on unpaved roads, to the small village of Nokaneng, on the western periphery of the Kalahari Desert.

By this point in their journey, it was July in the Southern Hemisphere, but the winter precipitation that had flooded the Okavango Delta was nowhere in sight at the edge of the Kalahari. The newlyweds created a staging ground in Nokaneng, leaving behind sufficient petrol for future travels, and then set out due west across the desert, toward the Namibian border. In the end, it took them eight hours to drive sixty miles through arid terrain.[2]

It was a grueling voyage, and in a sense, it was also a journey back in time. At the end of their eight-hour pilgrimage lay one of few regions of the Kalahari with sufficient water to support small communities of human beings, thanks to the nine waterholes spread out across an otherwise barren, flat landscape roughly 100,000 square miles in size. This more hospitable stretch of the Kalahari was sometimes referred to as the Dobe region, after the name of one of its waterholes. Howell and Lee had made their arduous journey because the Dobe region was the home of the !Kung people, a hunter-gatherer society that had been almost miraculously isolated from all the conventions and technology of modern life. The !Kung had managed to survive the preceding bloody centuries with almost no contact with other African societies and their European colonizers. They were protected, as Howell would later observe, "by the simple fact that none of the stronger peoples of southern Africa wanted to take their territory away from them, or even share it."[3]

Like many surviving hunter-gatherer societies around the world, the !Kung people offered Western anthropologists a provocative hint of the ancestral environment that had shaped most of the evolutionary history of *Homo sapiens*, before the agricultural revolution first arrived roughly ten thousand years ago. Lee had already visited the !Kung society several times before 1967 to study their social organization, their food production techniques, and their strategies for managing and sharing resources within the community. Lee's research had been instrumental in proposing a new way of thinking about hunter-gatherer communities, one that undermined the long-standing view, most famously captured in Thomas Hobbes's description of the "state of nature" as "solitary, poor, nasty, brutish and short." Observed up close, the !Kung did not appear to be struggling to get by, as Hobbes had assumed, in an arduous existence on the edge of starvation. Despite the paucity of natural resources around them, they seemed instead to enjoy a remarkably high standard of living, working less than twenty hours a week to support their nutritional needs. Based on similar research conducted on hunter-gatherer cultures in the Pacific, the anthropologist Marshall Sahlins had recently proposed a term for this reimagined model of early human social organization: the "original affluent society." The !Kung and their equivalent did not represent some impoverished past, woefully deprived of all the advancements of modern technology. Instead, Sahlins argued, "The world's most 'primitive' people have few possessions, but they are not poor."[4] Measured by the usual conventions of Western civilization, the !Kung did indeed appear to be primitive: they lacked transistor radios and washing machines and multinational corporations. But measured by more elemental standards—food, family, human connection, leisure—they seemed far more competitive with the industrialized world than conventional wisdom at the time had assumed.

It was another kind of measurement that had brought Nancy Howell halfway across the world to the Dobe region, perhaps the most elemental measure of a human life there is. The !Kung offered at least some meaningful evidence that could help determine if

Nancy Howell and her colleague Gakekgoshe conducting social
network research with the !Kung people, 1968
(Photo credit: Richard Lee)

early human existence was indeed "solitary, poor, nasty, brutish, and short." But as a demographer, Howell was particularly interested in the last of Hobbes's adjectives. How short were their lives exactly, compared to those of humans living in technologically advanced societies? How likely were they to live long enough to see their grandchildren? How likely were they to suffer the loss of a child, or die during childbirth? Affluence, after all, can be measured in leisure time, calorie intake, personal liberty—but surely one of the most important measures of an allegedly affluent society is how much life—and how little death—you experience as a member of that society.

Over the course of their three-year stay, Howell and Lee generated endless stacks of data: tracking kinship relations, pregnancies, calories consumed. But for Howell the most tantalizing—and elusive—number was one that has been the cornerstone of demography for its entire existence as a science: life expectancy at birth.

The number was elusive for several reasons. The !Kung kept no written records of their population histories; they had no census data to share with Howell, no mortality tables. Howell and Lee were only spending a few years among the !Kung, not nearly long enough to conduct a longitudinal study of the population, observing births and deaths over many decades. But the most confounding hurdle was the simple fact that the !Kung themselves had no idea how old they were, in part because their entire numerical system topped out at the number three. If you asked a member of the !Kung society what age they were, you got only blank stares. Age as a numerical concept simply didn't exist for them.

This was the challenge that Nancy Howell confronted as she and her husband set up camp in the Dobe in late July 1967. How do you compute life expectancy in a culture that doesn't bother to count years?

THE PRACTICE OF RECORDING the ages of a given culture's population is almost as old as writing itself. Archaeological evidence suggests that as far back as the fourth millennium BCE the Babylonians regularly conducted a census—probably for taxation purposes— that registered both overall population size and the age of individual citizens, capturing the data on clay tablets. But the concept of life expectancy is a relatively modern invention. Census data is a matter of facts: *this man is forty years old; this woman is fifty-five.* Life expectancy, on the other hand, is something else altogether: a prediction of future events based not on sorcery or anecdote or guesswork but rather on the sturdier foundation of statistics.

The first calculations of life expectancy were inspired by an unlikely source: a British haberdasher by the named of John Graunt, who conducted an elaborate study of London mortality reports in the early 1660s entirely as a hobby, publishing his findings in a 1662 pamphlet titled *Natural and Political Observations Mentioned in a following Index, and made upon the Bills of Mortality*. The fact that Graunt had no formal training as a demographer shouldn't surprise us; neither demography nor the actuarial sciences existed as formal disciplines back then. Indeed, Graunt's pamphlet is widely considered the founding document of both fields. Statistics and probability were themselves in their infancy during this period. (The word *statistics*, in fact, wouldn't be coined for more than a century; in Graunt's time, it was known as *political arithmetic*.) It remains something of a mystery, though, why Graunt himself decided to take up the problem of calculating life expectancy. One motivation was clearly altruistic: Graunt suspected that a close analysis of the city's mortality reports might alert the authorities to outbreaks of bubonic plague, allowing them to establish quarantines and other crude public health interventions. Thanks to this idea, Graunt is also considered one of the founders of epidemiology, though his pamphlet did little to arrest the devastating plague that erupted three years later in 1665, famously recounted in Samuel Pepys's diary and in Daniel Defoe's semifictional *A Journal of the Plague Year*.

While Graunt had been trained as a haberdasher, by the time he took up his amateur interest in demography, he had become a successful and well-connected businessman, serving as an officer in an international trading firm known as the The Drapers' Company. He served on several city councils and socialized with Pepys as well as with a polymath surgeon and musician named William Petty, who would go on to write a number of influential books on political economy and statistics, including one called *Political Arithmetic*. (A small subset of scholars of this period actually believe that

Petty wrote the *Natural and Political Observations of the Bills of Mortality*, and not Graunt.) In the introduction, Graunt claims the original idea for the project occurred to him after many years of observing the way Londoners read the Bills of Mortality, the weekly catalog of citywide deaths that had been dutifully compiled and published by a guild of parish clerks since the early 1600s. The readers, Graunt observed, "made little other use of them, than to look at the foot, how the Burials increased, or decreased; and, among the Casualties, what had happened rare, and extraordinary in the week current: so as they might take the same as a Text to talk upon in the next Company."[5] Londoners would scan the Bills for the headlines. (How many dead this week? Any interesting new diseases on the march?) If something caught their eye, they might pass the information off casually to a friend over a pint. But no one bothered to look at the Bills systemically, as data that might suggest a wider truth beyond the random fluctuations of each week's death toll.

Graunt's work proposed a radical break from that history of neglect. He would use the data not as fodder for idle gossip but as a way of testing hypotheses about the overall health of London's population, and as a way of perceiving long term trends in that community. His investigation began with an informal perusal of a handful of Bills of Mortality, which suggested a few "Conceits, Opinions, and Conjectures"—as Graunt would later put it—about the city's health. Inspired by that initial set of queries, he spent months visiting Parish Clerks Hall on Brode Lane, just north of Southwark Bridge, acquiring as many Bills of Mortality as he could for his research. After a painstaking tabulation of the data—assembled centuries before the invention of calculators, much less spreadsheets—Graunt produced about a dozen tables that formed the centerpiece of his pamphlet. He began with one of the core questions of modern epidemiology: What was the distribution of causes of death in the population? To answer this question, he drew up two tables, one displaying "Notorious Diseases" and the other "Casualties." Both tables echo the famous "Chinese encyclopedia" from Jorge Luis Borges, with an eclectic mix of categories that seems comical to the modern eye. The "Notorious Diseases" table reads as follows:[6]

Apoplex	1306
Cut of the Stone	38
Falling Sickness	74
Dead in the Streets	243
Gout	134
Head-ach	51
Jaundice	998
Lethargy	67
Leprosie	6
Lunatick	158
Overlaid and Starved	529
Palsie	423
Rupture	201
Stone and Strangury	863
Sciatica	5
Suddenly	454

The "Casualties" table featured a number of culprits that would be familiar to a contemporary demographer—he counted 86 murders, for instance—while other causes of death might raise an eyebrow: Graunt reported that 279 people in his survey died of "grief," while 26 were "frighted" to death.

The most crucial tabulations, however, involved what Graunt called the "acute and epidemical diseases": smallpox, plague, measles, and tuberculosis, which Graunt called consumption, using the terminology of the day. Calculating the total number of deaths over the period, and then breaking down that total into its component parts, allowed Graunt—for the first time—to propose an answer to the question, How *likely* were you to die from a particular cause? The Bills of Mortality were simply an inventory of deaths, facts without meaning beyond the individual, human tragedy of the life lost. Graunt's tables took the facts and transformed them into probabilities, which gave the authorities an actionable overview of what the major threats to public health were, insights that would allow them to combat those threats and prioritize between them more effectively.

But the most revolutionary statistical technique that Graunt introduced appeared in a chapter titled "Of the Number of Inhabitants." Graunt began the chapter referencing multiple conversations he had conducted with "men of great experience in this city" who suggested that the total population of the city must be in the millions. Graunt correctly perceived from his study of the mortality reports that the figure must be greatly exaggerated. (A city of 2 million people would have had far more deaths than were recorded in the Bills.) Through a number of roundabout calculations, Graunt proposed a much lower number: 384,000. Graunt himself thought the number had been determined "perhaps too much at random," but the calculation has held up well since he first published it: modern historians estimate London's population during this period to have been somewhere in the range of 400,000.[7]

Armed with this crucial denominator—total population—Graunt was then able to examine another key element of the Bills of Mortality in a new light: age at death. He divided the overall pool of recorded deaths into nine separate tranches: those that died before their sixth birthday; those that died between their sixth and sixteenth birthdays; between their sixteenth and twenty-sixth birthday; and so on all the way up to eighty-six. With the deaths segmented in this fashion, Graunt was able to calculate the distribution of deaths in the population by age. For every hundred Londoners born, Graunt reported, thirty-six of them would die before their sixth birthday. In modern terminology, we would call that a childhood mortality rate of 36 percent.

The entirety of Graunt's "life table" was sobering. Less than half of London's population survived past adolescence; fewer than 6 percent made it to their sixties. Graunt did not manage to take the next step and reduce his life table to the single number we now use as perhaps the most fundamental measure of public health: life expectancy at birth. But we can calculate it based on the data Graunt did assemble in the table. By Graunt's account, the life expectancy of a child born in London in the mid-1660s was only seventeen and a half years.

WHEN NANCY HOWELL ARRIVED in the Dobe region in the middle of 1967 and began her investigation into the health and life spans of the !Kung people, she possessed several crucial advantages over John Graunt in attempting such a study. She had three hundred years of advances in statistics and demography at her disposal. Since

Graunt's time, demographers had developed extensive tools to calculate not only life expectancy at birth, but also, crucially, life expectancy at other ages as well. Howell had more than just conceptual
tools, however. She had data entry systems and calculators to crunch
the numbers; she had cameras to photograph the !Kung to help
identify them for the records and connect them to studies that had
been completed earlier in the decade. She had tape recorders to
capture her interviews with the !Kung. She would even ultimately
develop a software program—called AMBUSH—to simulate the
fluctuations in the !Kung population over time.

Yet with all these assets, Howell still confronted the onerous
challenge of conducting a functional census: the obstinate fact that
the !Kung had no concept of age as a numerical category measured
in years. Even making rough estimates of age based on visual appearance was challenging. Many of the !Kung who would turn out
to be in their sixties appeared to Western eyes to be much younger
than they were, thanks to the active lifestyle and distinct diet of the
hunter-gatherer society. In addition, there were no Bills of Mortality to consult, no written records at all, in fact. Somehow Howell
would have to do the work of the parish clerks, in a culture that did
not find it necessary to use numbers higher than three.

As Howell contemplated the task ahead in September 1967, the
prospects seemed daunting, made even more intimidating by the
climate of the Kalahari during that part of the year. The rains had
stopped months before; daytime temperatures regularly climbed above
110 degrees; and most temporary water sources had evaporated in
the heat.

Howell managed to turn the hostile conditions of the Kalahari
dry season to her advantage. With temporary water supplies unavailable until the rains began again near the end of the year, the !Kung

people congregated around the main water holes that defined the region. Howell and her husband became frequent guests in the small villages that surrounded each water hole. They would arrive with a scale and a height rod and a bag of tobacco. The two scholars would dole out the tobacco and put on an informal measuring party, where members of the community would have their weight and height recorded. Howell would later write that the !Kung looked forward to the visits, "since they provided them with supplies of tobacco and a break in routine to sit in the shade for a few hours joking and watching others being measured. They also provided a convenient opportunity to collect a great deal of casual information and news about the groups."[8]

As successful as the measuring parties were in calculating weight and height, the unit of measure that Howell was most interested in—age—was not as readily captured. In the end, the key stratagem that enabled Howell to calculate a reasonably accurate assessment of each !Kung's age in years was grammatical, not numerical. While the !Kung did not count age in years, they had a finely tuned sense of *relative* age. They were fully aware of which members of the community were older than they were, and which were younger.[9] That age distinction was reflected in their spoken language: just as many Indo-European languages, such as French and Spanish, differentiate between formal and informal relationships in direct address (the difference, in French, between "vous" and "tu"), the !Kung language had a comparable grammatical category that differentiated between an older person and a younger person. In effect, when a member of the !Kung uttered a statement along the lines of "Will you help me prepare this meal?" the actual meaning of the question would be: "Will you, younger person, help me prepare this meal?"

That tiny syntactical distinction ended up giving Howell enough clues to eventually crack the case of !Kung life expectancy. Lee had already assembled a rough census for the Dobe population based on earlier visits to the region dating back to 1963. From his own observations of births during that period, he could establish the ages of younger children with meaningful accuracy. A child whom Lee had seen as a toddler in, say, 1963 could be reliably pegged as a six- or seven-year-old in 1967. That gave Howell a floor for building her investigation. She could listen to that six-year-old in casual conversations with his friends and note which ones of those friends was greeted with the younger-than-me form of address, and which ones received the older-than-me version. She supplemented that data by directly interviewing 165 !Kung women who were of childbearing age or beyond. In those interviews Howell recorded detailed fertility histories: pregnancies, miscarriages, abortions, stillbirths, successful births, and more. Those were events that could be mapped onto years as well because they were often spaced in one- or two-year intervals. A mother might report to Howell that she'd had a miscarriage two years ago, and then two years before that her daughter was born, making her daughter four years old. By tracking this web of familial and social connections, Howell was able to build a kind of hierarchy: a ranked list of the population organized by age. The exact ages grew blurrier as the population thinned out at the older end of the spectrum: if there were only two people in their seventies, it was hard to tell exactly how old either of them was, only that one was older than the other. But it was close enough to get the general picture of !Kung life expectancy.

In her analysis, Howell saw evidence that !Kung life expectancy at birth had improved over the preceding centuries, perhaps as an effect of some elements of modern health systems infiltrating their hunter-gatherer culture. She ultimately came to argue that a child born into the !Kung society in the late 1960s could expect to live, on average, thirty-five years, while the older generations experienced a life expectancy of thirty years. That seems short by our modern standards, but in fact many of the !Kung enjoyed life spans that would be considered long even in developed countries during the late 1960s. In one of her books, Howell describes a !Kung elder named Kase Tsi!xoi, who was eighty-two when Howell interviewed and photographed him in 1968.[10] He was still sturdy enough to gather his own food and travel by foot for long distances. When Howell first encountered him, he was in the process of constructing a hut for himself in a new settlement.

The main factor keeping !Kung life expectancy low was the relatively high rate of infant and childhood mortality, which turned out to be not all that different from the mortality rates Graunt had observed in London three hundred years earlier. Two out of ten children failed to survive the first months after birth, and another 10 percent died before they reached their tenth birthday. There were far more grandparents and great-grandparents than you might have initially expected in a society with a life expectancy of thirty-five. If you made it past adolescence in the !Kung culture, you had a reasonable chance of making it to your sixties and beyond. The problem was that making it to your sixties meant that you had most likely experienced the deaths of multiple children and grandchildren over the course of that life. For the !Kung, it was not all that difficult to grow old, once you made it through the crucible of childhood.

JOHN GRAUNT'S PAMPHLET ANALYZING the Bills of Mortality was an immediate success. The haberdasher was invited to join the prestigious Royal Society, and copies of his essay circulated widely among both mathematically minded Europeans and nascent public health officials. (Inspired by Graunt's statistical analysis, Paris introduced its own version of the Mortality Bills in 1697.) Probability theory was in its infancy in the middle of the seventeenth century; the idea of using math to ascertain the likelihood of a certain event happening was a genuinely novel concept when Graunt first started surveying the parish clerks' data sets. Ironically, while Graunt was wrestling with existential questions about life and death, almost all of the important work on probability up until that point had been directed toward a far more frivolous question: how to win at games of chance, such as dice or cards. Graunt's tables suggested a new use for these emerging mathematical tools: if you could accurately assess the risks and opportunities of dice games, could you use those tools to do the same with the game of life?

The first genuine assessment of life expectancy appeared in a series of letters written in 1669 between the Dutch polymath Christiaan Huygens and his brother Lodewijk. Christiaan was one of the most influential and brilliant scientists of his time. As an astronomer he studied the rings of Saturn and made the first observations of Saturn's moon, Titan. He proposed the wave theory of light and invented the pendulum clock. He had also published a seminal sixteen-page treatise on probability theory, titled *De ratiociniis in ludo aleae*, Latin for "On Reckoning at games of chance," which introduced the crucial concept of "expected gain" to the field, now the cornerstone principle behind every casino business in the world. Based on this work, the president of the Royal Society had sent Christiaan a copy of Graunt's paper shortly after it was published, but it was the master scientist's brother Lodewijk who first proposed the life expectancy calculation.

Lodewijk's interest in the problem had financial roots. He had recognized that having a mathematically sound assessment of life expectancy would enable the fledgling insurance industry to price life annuities more effectively. A close cousin to pensions, life annuities are the opposite of traditional life insurance policies: annuities are paid out in regular installments as long as you live. From the insurers' purely mercenary perspective, a customer who dies young is more profitable than a customer who lives longer than expected. (The incentives are reversed in normal life insurance.) But establishing the price for both kinds of insurance depended on being able to measure expected life. You'd want to set the price of a life annuity a great deal higher in a society where the average person lived to be sixty as opposed to thirty-five or seventeen. And it would be particularly useful to be able to calculate not just overall life expectancy at birth, but also expected life based on a given person's age. How much more should an insurance company charge a twenty-year-old buying an annuity than a forty-year-old doing the same?

In a letter dated August 22, 1669, Lodewijk wrote to his brother of a strange hobby he had taken up over the past few weeks. "With regard to age," he wrote, "I have made a table these days past of the time that remains to live, for some persons of all sorts of ages." He had based his table on the original data set that Graunt had assembled in his pamphlet. In the letter, Lodewijk's pride in his work is evident. "The consequences which result from it are very pleasing and . . . can be useful for the compositions of life annuities." He mentioned one finding that was certain to get his brother's attention: "According to my calculation," he explained, "you will live until about the age of 56 and a half. And I until 55."[11]

Christiaan wrote back with some suggested modifications to his brother's math, and even sketched an ingenious graph representing

Graunt's data, the first known instance of what is now called a continuous survival function. Reading it now, it is impossible not to hear the intonations of sibling rivalry in the exchange, with Lodewijk no doubt straining to impress his overachieving brother, and Christiaan subtly undercutting his brother's achievement with his corrections. (It is equally impossible not to be amazed by the kinds of activities that the Huygens brothers appeared to entertain themselves with in their spare time.) The letters that passed between the brothers in the late summer of 1669 did not initially receive the same acclaim from august bodies like the Royal Society. But today we appreciate them as a transformative moment in the history of that most ancient of questions: *How long do I have to live?*

LODEWIJK HUYGENS TURNED OUT to be too pessimistic in his projections. Christiaan lived ten years longer than Lodewijk's calculation had predicted; Lodewijk himself lived to sixty-eight. But these were probabilities, not prophesies. Lodewijk's calculation—and the concept of life expectancy that emerged from it—distilled the teeming chaos of thousands of individual lives into a stable average. That analysis couldn't tell you how long you would actually live, but it could tell you how long you might reasonably *expect* to live, given the patterns of life and death in your surrounding community. And it suggested something equally important: a way of measuring the overall health of that community. For the first time it was possible to compare the overall health records of two societies or to track changes in a single community over time.

John Graunt's tables, in their own way, were too pessimistic as well. During the 1970s, a historical demographer named Anthony Wrigley organized a massive database of British parish records dating back to the middle of the sixteenth century; from those archives, Wrigley and his collaborators calculated the country's life expectancy rates from the end of the Renaissance era to the middle of the Industrial Revolution. Wrigley's analysis revealed that life expectancy at birth in London during the seventeenth century was just under thirty-five years.[12] (During outbreaks of the plague—like the particularly lethal outbreak of 1665–66—life expectancy would have briefly plunged closer to the seventeen years that Graunt's tables suggested.) Nancy Howell's analysis of hunter-gatherer life spans, on the other hand, has generally been supported by subsequent research. A number of scholars have analyzed fossils from preagricultural human settlements, estimating age through the presence of deciduous and permanent teeth in the skeletons of humans who died before the age of fifteen, and analyzing bone decay and other clues to assess the age at death of older members of the community. Between studies like Howell's, which examine existing hunter-gatherer tribes, and archaeological forensics that examine ancient human fossils, we now believe that our hunter-gatherer ancestors generally saw life expectancies somewhere between thirty and thirty-five years, and suffered childhood mortality rates upward of 30 percent.

Graunt and Huygens couldn't have known it at the time, but their first estimates of average life expectancy revealed something profound not just about European culture at the edge of the Enlightenment. They also revealed something profound about the entire ten-thousand-year run of human civilization, something that wouldn't be fully grasped until researchers like Nancy Howell began plotting the life expectancy of hunter-gatherer communities in the second half of the twentieth century. Our Paleolithic ancestors would have been confounded or mesmerized by the achievements of the civilization that John Graunt was born into: cities of four hundred thousand people, sharing news and information via printing presses, calculating mortality rates and financial transactions with alphanumeric codes, engineering palaces, bridges, and cathedrals—all the spectacular triumphs of postagricultural man. But despite those triumphs, the answer to the existential question—*How long do I have to live?*—would have been remarkably familiar to a hunter-gatherer transported to Graunt's London. The average person lived to his or her early thirties, but a meaningful portion of the population lived far beyond that. (Graunt himself died at fifty-three.) And almost a third of the population—in both hunter-gatherer societies and seventeenth-century London—died before they reached adulthood.

Thomas Hobbes had published his "nasty, brutish, and short" dismissal of the state of nature just a few years before Graunt started dabbling with the Bills of Mortality. But the revolution in demography and statistics that Graunt triggered—the one that ultimately led Nancy Howell to spend those years with the !Kung in the late 1960s—would eventually make it clear that Hobbes had it wrong with at least one of those famous three adjectives. Whatever you may think of preagricultural humans being nasty and brutish, their lives were certainly not short by the standard of Hobbes's era.

This extended view of *Homo sapiens* health over time offered a sobering prospect: despite all our achievements, we had remained trapped beneath the long ceiling of thirty-five years of life expectancy, with a third of all children dying before adulthood. Human beings had spent ten thousand years inventing agriculture, gunpowder, double-entry accounting, perspective in painting—but these undeniable advances in collective human knowledge had failed to move the needle in one critical area. Despite all those accomplishments, we were no better at warding off death.

FOR A CENTURY that followed the publication of Graunt's pamphlet, the health of European populations continued to follow the pattern that had been in place for millennia, bobbing up and down around a median of thirty-five years, propped up here by an unusually bountiful harvest, pushed down there by a deadly outbreak of smallpox or a harsh winter. Globally, life expectancy almost certainly declined, thanks to the growth of the slave trade and the catastrophic impact of European diseases imported to the Americas. But in Europe itself, there were no directional trends in the data, just seemingly random fluctuations around a life expectancy ceiling that had been in place since the Paleolithic era.

The first hint that this ceiling might be vulnerable appeared in England during the middle decades of the eighteenth century, just as the twin engines of the Enlightenment and industrialization began to power up. The change was subtle at first, largely imperceptible to contemporary observers, even to those experiencing the change itself. In fact, the change was not properly documented until the 1960s, when a demographic historian named T. H. Hollingsworth began analyzing the exacting records of births and deaths maintained by the College of Heralds and the publishers Burke's and Debrett's. As a measure of the overall population, these records were far less expansive than Graunt's, only tracking the lives of a tiny—though particularly interesting—subset of the British population: the aristocratic class of English "peers." Hollingsworth uncovered data on every duke, marquess, earl, viscount, and baron—

and their children—from the late 1500s all the way to the 1930s. When all that data had been assembled into a chart of life expectancy trends, a startling pattern emerged.[13] After two centuries of stasis, right around 1750 the average life expectancy of a British aristocrat began to increase at a steady rate, year after year, creating a measurable gap between the elites and the rest of the population. By the 1770s, British peers were living on average into their mid-forties. They crossed the threshold of the fifty-year mark at the dawn of the nineteenth century, and by the middle of Victoria's reign they were approaching a life expectancy at birth of sixty.

BRITISH LIFE EXPECTANCY AT BIRTH, 1720-1840

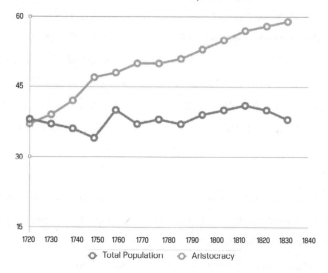

British life expectancy at birth, 1720–1840

At a time when the world's population was numbered in the hundreds of millions, those British peers constituted a vanishingly small proportion of humanity. But the demographic transformation they experienced turned out to be a glimpse of the future. As far as we know, this was the first time in history when life expectancy started to increase at a steady and sustained pace, among a meaningful population of humans. The endless bobbing of the previous ten thousand years had taken on a new shape: a straight line, slanting upward.[14]

The takeoff in life expectancy among British peers was notable for another reason. It marked the beginning of a pattern that would become an inescapable reality for much of the world in the subsequent centuries: a measurable gap in health outcomes between different societies, or between different socioeconomic groups within the same society. In John Graunt's time, it didn't matter whether you were a baron or a haberdasher or a hunter-gatherer: your life expectancy at birth was going to be in the mid-thirties. If you happened to be born into an elite family in a major metropolitan center, you would have the opportunity to enjoy many of the trappings of civilization: fine art, comfortable housing, plentiful food. But all that wealth would have given you no advantage whatsoever over your less affluent contemporaries at the elemental task of keeping yourself and your family alive. (Strangely enough, it may have actually given you a slight disadvantage—a paradox that we will explore shortly.) There were great inequalities in health outcomes person to person: many died at eight days, while some lived to eighty years. But the inequalities of life span—sometimes called gradients—did not arise between large social groups. That would change by the second half of the eighteenth century. Health inequalities began to appear alongside wealth inequalities, a trend that first became visible among the British peerage, with its life expectancy at birth climbing thirty years in a century, while the working classes languished in conditions that would have been right at home in Graunt's tables from 1662.

By the second half of the nineteenth century, both patterns

would spread beyond that small advance guard on the British Isles and make their way around the world. The straight line, slanting upward, came to describe the life expectancies of ordinary Europeans and North Americans, not just the aristocrats. By the first decade of the twentieth century, overall life expectancies in England and the United States had passed fifty years. Millions of people in industrialized nations found themselves in a genuinely new cycle of positive health trends, finally breaking through the ceiling that had limited *Homo sapiens* for the life of the species. But at the same time, that great escape, as the historian and Nobel laureate Angus Deaton has called it, opened up a tragic gradient in outcomes between the industrialized countries and the rest of the world. Exploited by Western imperialism, devastated by European diseases, unaided by early public health institutions taking shape in Europe and North America, societies in the developing world not only failed to join the upward ascent of their industrialized peers—in many cases, they went backward. Life expectancies in parts of Africa, India, and South America dropped below thirty years. "It is possible that the deprivation in childhood of Indians born around midcentury was as severe as that of any large group in history," Deaton writes, "all the way back to the Neolithic revolution and the hunter-gatherers that preceded them."[15] The great lottery of life—where you were born, what socioeconomic group you were born into—now played a major role in determining whether you survived the perilous years of early childhood, or lived long enough to meet your grandchildren. By the first years of the twentieth century, undeniable progress in health outcomes had been achieved in the wealthy parts of the world. But was that progress sustainable? And could the fruits of that progress be shared with the rest of the world?

The answers to those questions depended, in part, on understanding what had driven the first upward trajectory of the great escape. Why were Westerners living longer? Why were their children no longer dying at such catastrophic rates? These questions had both historical and practical significance. If we could identify what was improving health outcomes in Europe and the United States, presumably those interventions could then be spread to the rest of the world. The ultimate explanation for the first sustained extension of life expectancy proved to be less straightforward than you might expect. It seemed logical to attribute a society-wide improvement in health to the healthcare establishment of the day: doctors, hospitals, medicine. But that assumption—as self-evident as it might seem—turned out to be incorrect. If medicine was doing anything during this period, it was shortening lives, not extending them.

IN THE LATE SUMMER of 1788, the English monarch George III and his retinue traveled back to the royal estate at Kew in the suburbs of London, after spending two months "taking the waters" in Cheltenham, the king's first genuine vacation in thirty years. The idyll had been conceived as a health intervention, after George complained of painful spasms that lasted as long as eight hours. The countryside did appear to have a positive effect on the king's condition, but shortly after the return to London, he began experiencing even more painful attacks. His doctor, Sir George Baker, noted in his diary, "I found the king sitting up in his bed, his body being bent forward. He complained of a very acute pain in the pit of the stomach, shooting to the back and sides and making respiration difficult."[16] Baker prescribed castor oil and senna, two common laxatives, but then feared the dose had been too extreme and attempted to counteract it with a tincture of the opiate laudanum. The medicines had little effect. Within days, the planned return to Windsor Castle had been postponed and the king's normal schedule of appearances canceled.

George's spasms in October 1788 would turn out to be the first wave of one of history's most famous illnesses, one more noted for its psychological symptoms than for its physical ones. Thanks to some brilliant modern forensic sleuthing, the story of Mad King George also gives us clear evidence of just how incompetent medicine was during the first stirrings of the great escape. For several months, the king descended into a state of general derangement: foaming at the mouth, erupting into fits of violent rage, talking in endless sentences with little logic or coherence. The episode sparked a constitutional crisis and was later dramatized in the play and feature film, *The Madness of King George*. Interestingly, the first symptom of true mental disorder that George displayed was a volcanic outburst directed toward Baker, complaining about the medicines the doctor had been prescribing him. In his journal, Baker recounted his shock at the king's demeanor: "The look of his eyes, the tone of his voice, every gesture and his whole deportment, represented a person in the most furious passion of anger. One medicine had been too powerful; another had only teased him without effect. The importation of senna ought to be prohibited, and he would give orders that in future it shall never be given to any of the royal family." The tirade lasted three hours. When it had finally subsided, Baker wrote to William Pitt, the prime minister, and reported that the king was in an "agitation of spirits bordering on delirium."[17]

Medical historians have long debated the cause of King George's illness. Since the late 1960s, a consensus has emerged that George suffered from a hereditary condition known as variegate porphyria, which can cause abdominal pain as well as anxiety and hallucinations. (The genetic disorder is in fact known to be prevalent in the royal families of Europe—yet another argument for not marrying your close relatives.) Other scholars have argued that the king's unusual behavior during the winter of 1788 was the result of a bipolar disorder. But recent forensic studies suggest that George's outrage at his physician's treatment may have had some justification. In the early 2000s, a team of scientists led by a Cambridge metabolic physician named Timothy Cox analyzed a lock of George's hair that had been stored in the archives of the Wellcome Trust for almost a century. Cox and his colleagues knew that earlier attempts had failed to extract DNA from the hair samples to test for the presence of a gene known as PPOX. (Porphyria is caused by a malfunctioning PPOX gene.) Instead, the researchers analyzed the strands for the presence of heavy metals that could have exacerbated the king's illness. The results were astounding: arsenic levels in the hair were *seventeen* times higher than the standard threshold for arsenic poisoning. Analyzing the official reports of the king's physicians from the period in question, Cox and his colleagues found that the principal compound delivered to George was a then-popular treatment known as emetic tarter, which contained somewhere between 2 and 5 percent arsenic. Assuming the dosages recorded in the physician's reports were accurate ones, King George's "treatment" for his delirium and abdominal pain appears to have been chronic arsenic poisoning.[18]

Given the hereditary issues with porphyria in his family, it should not surprise us that George III experienced mental health issues during his reign. Much more surprising is that he survived the attempts to cure him.

. . .

IN THE EARLY 1960S, when T. H. Hollingsworth published his analysis of the life expectancies of the British peerage, the demographic historian gave us the initial glimpse of the great escape in its embryonic form—all those dukes and barons surviving infancy and living into their sixties and beyond, presaging the health trends that would envelop the entire globe two centuries later. But there was a curious footnote to Hollingsworth's study. Take a look at the original chart of the first great escape, this time with the preceding two centuries included.[19]

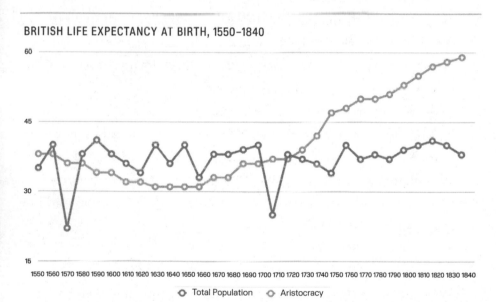

BRITISH LIFE EXPECTANCY AT BIRTH, 1550–1840

British life expectancy at birth, 1550–1840

In the century before the elites began living longer than the rest of the population, the average peer actually possessed a slightly *lower* life expectancy than the average commoner. The gap was not nearly as pronounced as the one that quickly developed late in the 1700s—only a few years difference separated the two groups—but the gap was consistent and statistically meaningful. It was a mysterious finding. All the advantages of affluence, social status, education resulted in a net *disadvantage* where life expectancy was concerned. Something was killing the aristocracy of England at a higher rate than the commoners. But what was it?

The most likely explanation for this strange gap is a counterintuitive one: the British peers had better access to health care than the rest of the population. They could afford consultations with as many doctors and surgeons and druggists as they liked. Because the state of medicine was so abysmal, those interventions actually did more harm than good. If you were unlucky enough to come down with the flu, or be born with a hereditary disorder that caused porphyria, you were better off avoiding doctors altogether and letting your body's immune system work to heal you rather than seeking out the phony cures of arsenic or leeches.

These quack cures were hardly limited to the English peerage. Consider the final hours of King George's great nemesis, George Washington, recounted in William Rosen's masterful history of the invention of antibiotics, a description that is practically indistinguishable from a torture manual:

> By the time the sun had risen, Washington's overseer, George Rawlins . . . had opened a vein in Washington's arm from which he drained approximately twelve ounces of his employer's blood. Over the course of the next ten hours, two other doctors—Dr. James Craik and Dr. Elisha Dick— bled Washington four more times, extracting as much as

one hundred additional ounces. Removing at least 60 percent of their patient's total blood supply was only one of
the curative tactics used by Washington's doctors. The former president's neck was coated with a paste composed of
wax and beef fat mixed with an irritant made from the
secretions of dried beetles, one powerful enough to raise
blisters, which were then opened and drained, apparently
in the belief that it would remove the disease-causing poi-
sons. He gargled a mixture of molasses, vinegar, and butter;
his legs and feet were covered with a poultice made from
wheat bran; he was given an enema; and, just to be on the
safe side, his doctors gave Washington a dose of calomel—
mercurous chloride—as a purgative. Unsurprisingly, none of
these therapeutic efforts worked.[20]

Scholars now call this period the age of heroic medicine—full of
grand schemes and bold interventions that clearly did more harm
than good. Some of those interventions proved to be mere folly,
like those pastes and poultices applied to Washington on his death-
bed. But many of them warranted a malpractice suit. Bleeding sick
people was just as likely to hasten their deaths. Mercury and arsenic
can kill you or push you over the edge into clinical insanity. As we
will see, the age of heroic medicine survived far longer than we might
now expect. As late as the onset of World War I, William Osler, the
founder of Johns Hopkins, advocated for bloodletting as a primary
intervention for military men who came down with influenza and
other illnesses: "To bleed at the very onset in robust, healthy individuals in whom the disease sets in with great intensity and high
fever is, I believe, a good practice."[21]

. . .

THE FIRST SCHOLAR TO challenge the link between industrial-age medical science and life expectancy was a British-Canadian polymath named Thomas McKeown. In the late 1930s, as war was breaking out across Europe, McKeown moved to London for medical school after a stint at Oxford on a Rhodes Scholarship. Years later he would describe the experience as a turning point in his intellectual development. While observing physicians doing their rounds at a hospital, McKeown noticed a strange absence in their interactions with patients, and in their discussions with their peers. The doctors took vital signs, listened intently to descriptions of symptoms, and doled out their advice for treatment. But according to McKeown, they rarely wrestled with the question of "whether the prescribed treatment was of any value to the patient." Though Mc-Keown would eventually graduate in 1942 with a degree in surgery, his skepticism about the interventions performed in the hospital only grew stronger during his medical school years. He later wrote of that period: "I adopted the practice of asking myself at the bedside whether we were making anyone any wiser or any better, and soon came to the conclusion that most of the time we were not."[22]

At the end of the war, McKeown was offered a tantalizing academic position at the University of Birmingham: a newly endowed chair in "social medicine." He would occupy that position for the rest of his professional life. In the early 1950s, he began a research project that built on his intuitive insights from doing rounds as a med student, a project that would culminate more than two decades later in the form of a book called *The Modern Rise of Population*, one of the most controversial and influential studies of demographic change ever published. More than forty years after it first appeared, the book's argument—now known as the McKeown thesis—is still stirring debate.

The Modern Rise of Population proposed answers to two crucial questions about the last two centuries. First, was the overall growth in population during that period the result of increased fertility or decreased mortality? For this question, McKeown made a definitive case that the primary cause was not people having more babies, but rather the existing babies going on to live much longer lives. In England during the second half of the nineteenth century, birth rates dropped about 30 percent, even as the overall population doubled in size. But that fact suggested a thornier question: What exactly was keeping those babies alive? What drove the great escape in life expectancy that began in the last decades of the century?

Up until McKeown began publishing his findings, the answer to that question traditionally assumed that improvements in medicine had played a crucial role. The assumption was a natural one: if people were living longer, if they were not succumbing to disease at the same rates as their ancestors, it must be a sign that the medical professionals were getting better at their job. His medical school years had made McKeown naturally suspicious of this conventional wisdom, but when McKeown looked at the historical data, one fact jumped out at him: people stopped dying of diseases *before* doctors had working cures for them. In a passage in the book's opening pages, McKeown put that pattern at the very epicenter of his argument:

> In the period since the cause of death was first registered, a large majority of infectious deaths were due to the following diseases, which were also those associated mainly with the decline of mortality: tuberculosis, scarlet fever, measles, diphtheria and the intestinal infections. In all these diseases

it can be said without reservation that effective immu-
nization or therapy was unavailable before the twentieth
century.[23]

The data was unequivocal: the death toll from a disease like tuberculosis clearly declined over the final decades of the nineteenth
century, and into the first decades of the twentieth. And yet the
weapons against tuberculosis deployed by state-of-the art medicine
during that period were no more effective than the heroic treatments
given to the mad king George. (Though they may have caused less
actual damage to the patient.) Something had happened to cause
tuberculosis rates to decline in the human population of England.
And it wasn't the doctors. So what was it?

McKeown ultimately offered an alternate explanation: people
were living longer not because of medical interventions but because
of an overall improvement in the standard of living, thanks in large
part to agricultural innovations that put more food on the table.
As we will see, this part of McKeown's theory has since been chal-
lenged by more recent scholarship, but his diagnosis of the sorry
state of medicine has stood the test of time: Most historians now
believe that, in total, medical interventions had a limited effect on
overall life expectancy until the end of World War II. Whatever
positive effects were caused by genuinely useful knowledge or medicine the doctors had accumulated before that point were canceled
out by the lingering delusions of leeches and arsenic, all the ludi-
crous interventions of heroic medicine. Until the late nineteenth
century, they were also canceled out by the shockingly unhygenic
conditions of most hospitals and other medical environments. The
question of *why* such dubious practices took so long to be over-
thrown is a fascinating one. We will return to it in due time. But

the surprisingly long life of heroic medicine, and all its absurdities, should serve as a useful reminder that "Western" medicine—for all of its recent achievements—had a miserable track record for most of its existence. And in fact, the first intervention to have a meaningful *positive* impact on life expectancy didn't originate in the West at all.

2

THE CATALOGUE OF EVILS

VARIOLATION AND VACCINES

No one knows exactly when and where variolation was first practiced. Some accounts suggest it may have originated in the Indian subcontinent thousands of years ago. The historian Joseph Needham described an eleventh-century "Taoist hermit" from Szechuan who brought the technique to the royal court after a Chinese minister's son had died of smallpox.[1] The sixteenth-century Chinese pediatrician and medical writer Wan Quan makes reference to a technique whereby healthy children were deliberately exposed to variola minor, the less harmful cousin of smallpox. Whatever its origins, the historical record is clear that the practice had spread throughout China, India, and Persia by the 1600s. Like many great ideas in the canon, it may have been independently discovered multiple times in unconnected regions of the world.

The technique took a number of forms. The Chinese practitioners removed scabs from a recovering smallpox victim and ground them down into a fine powder that was then blown into the nostrils, where it was absorbed by the mucous membranes. In Turkey, the preferred technique involved making an incision in the arm either with a needle or a lancet and inserting a small amount of material extracted from variola minor pustules. No one at the time understood the biological mechanism that made variolation work, but the general principle was clear: exposing people to a lesser form of the disease caused most of them to be resistant to the disease in the future. Now, of course, we can describe variolation's magic in the language of immunology: by introducing a small quantity of the antigen—the infectious agent—variolation trained the antibodies of the immune system to recognize the threat and fight it off more effectively. The approach proposed a radical break with the methods of heroic medicine and all the other potions that healers had concocted over the centuries. The healer's role was not to supply some kind of magic element that would cure patients' ills; instead, the intervention merely unlocked powers that were latent in the patients themselves.

There is a pleasing symmetry in contemplating the discovery of variolation from the vantage point of twenty-first-century health care. The most exciting new breakthrough in medicine—immunotherapy, using the body's natural defenses to tackle chronic diseases like cancer or Alzheimer's—relies on the same basic mechanism that enabled the first great breakthrough in the history of extending life.

It should not surprise us that this first breakthrough was designed to protect us against smallpox. The disease had been a scourge since at least the age of the Great Pyramids. (The mummy of Ramses V has visible smallpox pustules on his face.) In Manchester and Dublin from 1650 to 1750, smallpox accounted for more than 15 percent of all deaths recorded. Young children were particularly vulnerable to the disease. In Sweden during the eighteenth century, 90 percent of smallpox mortality occurred in children under the age of ten.[2] The loss of all those young children delivered a tremendous blow to life expectancy during this period, but the emotional toll was undoubtedly worse. That most devastating of human experiences—the sudden death of a child—was an everyday reality for the entire population. Parents lived with the knowledge that at any moment their children could come down with a fever, followed by the telltale rash, and within a matter of days, their son or daughter would be dead—often followed in short order by their siblings. Compared to our modern experience, the whole notion of childhood was inverted. Today we think of children as emblems of vitality and resilience, the vigor of youth. As the Cambridge statistician David Spiegelhalter has observed, "Nobody in the history of humanity has been as safe as a contemporary primary school child." But in the age of smallpox, childhood was inextricably linked to sudden and catastrophic illness. Being a child was to forever be on the brink of death, and being a parent was to forever be haunted by that imminent threat.

The public toll of smallpox was just as severe. No virus has shaped the contours of world history as dramatically as variola major. Smallpox played a crucial role in the story of European imperialism, most notoriously in the epidemic that Cortez and his men brought to the Aztecs, ultimately destroying that ancient civilization. A Spanish priest traveling with Cortez conveys the scale of the devastation: "They died in heaps, like bedbugs. . . . In many places it happened that everyone in a house died, and, as it was impossible to bury the great number of dead, they pulled down the houses over them so that their homes become their tombs."[3] Western history, too, was transformed by variola major. The list of

European leaders felled by smallpox between 1600 and 1800 staggers the mind. During the outbreak of 1711 alone, smallpox killed the Holy Roman emperor Joseph I, three siblings of the future Holy Roman emperor Francis I, and the heir to the French throne, the grand dauphin Louis. Over the ensuing seventy years, the disease claimed King Louis I of Spain, Emperor Peter II of Russia, Louise Hippolyte, sovereign princess of Monaco, King Louis XV of France, and Maximilian III Joseph, elector of Bavaria. Add up all the major political figures assassinated around the world over the past two hundred years, and the total is still a fraction of those killed by the smallpox virus during those deadly centuries. Think of all the political realignments and insurrections and crises of succession that never would have happened had smallpox not so thoroughly infiltrated the ranks of the European elite.

One beneficiary—if that is the right word—of the smallpox assault was George III himself. After the Stuart queen Mary died childless of smallpox in 1694, the throne was set to pass to her sister Anne, who was herself in the middle of a stupendous effort to conceive an heir. Between 1684 and 1700, Anne somehow managed to become pregnant *eighteen* times, losing most of the pregnancies to miscarriages and stillbirths. Two daughters survived infancy but died before the age of two, most likely from smallpox infections. Only one child, William, Duke of Gloucester, survived early childhood. When he succumbed to smallpox at the age of eleven, the Stuart dynasty effectively ran out of heirs. Confronting a genuine crisis of succession, Parliament opted to throw the crown across the Channel, to the Hanover line, ultimately resulting in the coronation of George I, the grandfather of the mad king George. The Hanover line had numerous qualities in their favor—they were Protestants and descendants of King James, for starters—but one additional advantage they possessed was that young George had already been exposed to smallpox, which gives you some sense of how critical the disease was to political calculations in this period. Take smallpox out of the picture, and it is entirely likely that George I would never have crossed the English Channel, much less have found his way to Windsor Castle.

In the long run, though, the most significant aristocratic encounter with smallpox involved a well-bred and erudite young woman who contracted the disease in December 1715. Lady Mary Wortley Montagu was the daughter of the Duke of Kingston-upon-Hull, and wife of the grandson of the Earl of Sandwich. She was brilliant, witty, beautiful. As a teenager, she had written novellas; in her early twenties, she struck up a correspondence with the poet Alexander Pope. When she fell ill at the age of twenty five she was attended by two royal physicians, Dr. Mead and Dr. Garth, who treated her illness with a state-of-the-art regimen: she was bled every two days; she was fed purgatives and laxatives; and she received a regular dose of a medicine that was a mix of saltpeter—the key ingredient in gunpowder—and the ground-up remnants of a calcified mass extracted from the intestines of animals. The doctors prescribed beer and wine as her primary beverages.[4]

Miraculously, Lady Montagu survived her bout with smallpox—and the attempts to cure her—though she emerged from illness with her legendary beauty scarred by the telltale marks of the smallpox survivor. At the time, the news that Lady Montagu had triumphed over variola major seemed only meaningful to her close family and the aristocratic circles she traveled in; in the grand scheme of things, it was just one less death in the mortality reports. But Lady Montagu's survival would prove to be a major turning point in the battle against smallpox. She would turn out to be a crucial transmission vector, not for the disease itself, but rather for the one medically viable way of preventing it.

MARY MONTAGU PLAYED a dual role in the history of variolation: she was a connector and an evangelist. No doubt her encounter with smallpox had left her as emotionally scarred by the disease as she was physically scarred, and as a parent of young children, she would have been eager for any potential means of warding off the speckled monster, as variola major was sometimes called. But the same was invariably true of just about any parent living in Europe during the

Portrait of Lady Mary Wortley Montagu by Jonathan Richardson the Younger, 1725

early 1700s, at the height of smallpox's terrors. The factors that made Mary Montagu different from the rest were her keen powers of observation and her influence among the London elite—along with one crucial accident of history: shortly after her successful recovery from smallpox, her husband, Edward Wortley Montagu, was appointed ambassador to the Ottoman Empire. In 1716, after spending her entire life in London and the English countryside, Mary Montagu moved her growing family to Constantinople, living there for two years.

Montagu immersed herself in the culture of the city, visiting the legendary baths and learning Turkish in order to read the country's poets in their original tongue. She studied Turkish cooking and began dressing in the sumptuous caftans worn by affluent women in Constantinople, concealing her smallpox scars behind veils. She captured her experiences of living in a series of letters that were ultimately published after her death. The correspondence is noteworthy both for Montagu's discerning eye for the "Oriental" customs she observed on the streets of the Turkish capital, and for her literary talents as a travelogue writer. (The letters were also noteworthy for a number of appalling passages that defended the institution of slavery in the country, arguing that the Turkish slaves were in many cases treated better than British servants.) But the real historical significance of the letters lies in her description of a most unusual Turkish custom that she had observed first-hand:

> Apropos of distempers, I am going to tell you a thing that I am sure will make you wish yourself here. The Small Pox—so fatal and so general amongst us—is here rendered entirely harmless, by the invention of engrafting (which is the term they give it). There is a set of old women who

make it their business to perform the operation. Every au-
tumn in the month of September, when the great heat is
abated, people send to one another to know if any of their
family has a mind to have the smallpox.

They make parties for this purpose, and when they are
met (commonly fifteen or sixteen together), the old woman
comes with a nutshell full of the matter of the best sort of
smallpox and asks what veins you please to have opened.
She immediately rips open the one that you offer to her
with a large needle (which gives you no more pain than a
common scratch) and puts into the vein as much venom as
can lie upon the head of her needle, and after binds up the
little wound with a hollow bit of shell, and in this manner
opens four or five veins.[5]

Montagu wrote multiple variations of this description of variolation in letters back to her family and friends in London. In some
of those accounts, she mentioned that she had been so impressed
by the procedure that she intended to inoculate her son. While a
few scientific reports on Turkish variolation had been submitted to
the Royal Society, Mary Montagu's account proved to be the most
influential—in part because she did not merely describe the treatment, but actually introduced it into her immediate family. On
March 23, 1718, she sent a cursory note to her husband that announced: "The Boy was engrafted last Tuesday and is at this time
singing and playing, and very impatient for his supper. I pray God
my next may give as good an account of him." She added a note
about their infant daughter: "I cannot engraft the Girl; her Nurse
has not had the smallpox."[6]

Montagu had requested the procedure be executed by "an old Greek woman who had practiced this way a great many years," but according to the embassy physician, Charles Maitland, "her blunt and rusty needle . . . put the child to much torture." Maitland intervened and performed an additional inoculation, inserting the smallpox pus into the child's other arm via an incision he made with a lancet. After a few days of fever and an outbreak of pustules on both arms, Montagu's son made a full recovery. He would go on to live into his sixties, seemingly immune to smallpox for the rest of his life. He is considered the first British citizen to have been inoculated. His sister, who was successfully inoculated in 1721, after Montagu and her family had returned to London, was the first person to undergo the procedure on British soil.

Montagu knew she was taking a mortal risk by engrafting her children, though she had no way of calculating the magnitude of that risk precisely. We now believe that most variolation practices around the world resulted in a mortality rate that was roughly 2 percent. And a significant portion of the inoculated developed severe cases of smallpox that disfigured them for life. Imagine watching over your child suffering through the dark nights of a severe smallpox infection, wondering if he was soon to die because of a choice that you had made as his parent. But Montagu had seen enough of variola major to recognize that those dark possibilities were lesser threats than leaving your child vulnerable to smallpox in the wild. In a world where more than one in four children died before the age of ten— many of them killed by smallpox—the 2 percent chance of death by inoculation was in fact a risk worth taking.

Impressed by the successful inoculation of the Montagu children, Charles Maitland, who had returned to London as well, performed a trial inoculation of six prisoners at Newgate Prison, which also generated positive results. (The prisoners were promised a full pardon if they agreed to participate in the experiment.) Word spread quickly through the drawing rooms and palaces of aristocratic England: Mary Montagu had brought back a miracle cure from the Orient, one that finally promised an effective shield against the most terrifying threat of the era. In late 1722, the Princess of Wales directed Maitland to inoculate three of her children, including her son Frederick, the heir to the British throne. Frederick would survive his childhood untouched by smallpox, and while he died before ascending to the throne, he did live long enough to produce an heir: George William Frederick, who would eventually become King George III.

The royal inoculations proved to be a tipping point. Thanks in large part to Mary Montagu's original advocacy, variolation spread through the upper echelons of British society over the subsequent decades. A number of inoculations ended in tragedy, with well-born children dying from the medical intervention their parents had imposed on them. It remained a controversial procedure throughout the century; many of its practitioners worked outside the official medical establishment of the age. But the adoption of variolation by the British elite left an indelible mark in the history of human life expectancy: in that first upward spike that began to appear in the middle of the 1700s, as a whole generation of British peers survived their childhoods thanks to their increased levels of immunity to variola major.

THE STORY OF MARY MONTAGU and her unlikely role in the history of medicine gives us a helpful framework for thinking about the larger question of what drives genuine progress in society, progress that can be measured both by the decrease in child mortality—all those parents who did not suffer the loss of a child—and overall increases in life expectancy. What's striking about the story of Montagu's "discovery" of inoculation lies in how it departs from the conventional narrative of progress, where our lives are improved thanks to the discoveries of the heroic scientist, usually male and European, guided by the empirical methodologies developed in the Enlightenment, who finds his way to some world-changing idea through the sheer force of his intellect. In the long history of humanity's battle with dangerous viruses, the primary figure to play that role is Edward Jenner, the British doctor and scientist, now considered the "father of immunology" thanks to his development of the smallpox vaccine.

The story of Jenner's "eureka moment" is among the most familiar such narratives in the annals of scientific history, up there with Newton's apple and Franklin's kite-flying experiments. As a rural doctor, Jenner had observed a strange pattern in the distribution of smallpox cases in his community: milkmaids seemed less likely than the average resident to contract smallpox. Jenner hypothesized that the women had previously contracted a disease known as cowpox—a less virulent cousin of smallpox—thanks to their work routines; that exposure, he thought, had somehow granted them immunity to the more dangerous illness. On May 14, 1796, Jenner performed his now legendary experiment: scraping some pus from the cowpox blisters of a milkmaid, and inserting the material into the arms of an eight-year-old boy. The boy developed a light fever, but soon proved to be immune to smallpox. According to the standard account, Jenner's experiment constituted the first true vaccination, marking the beginning of a medical revolution that would save billions of lives over the subsequent centuries.

On one level the traditional focus on Jenner and his milkmaid epiphany is clearly warranted. May 14, 1796, does indeed constitute a watershed moment in the history of medicine, and in the ancient interaction between humans and microorganisms. But the spotlight on Jenner also keeps a crucial part of the action shrouded in darkness, distorting our perception of how these transformative health breakthroughs really happen. Jenner himself had been inoculated as a young child in 1757, and in his capacity as a local doctor, he regularly inoculated his patients. As a scientist and a doctor, Jenner had inherited a long-established principle that injecting smallpox-infected material subcutaneously could produce immunity. Without a lifelong familiarity with variolation, it is unlikely that Jenner would have hit upon the idea of injecting pus from a less virulent but related disease. As Jenner would later demonstrate, vaccination improved the mortality rates of the procedure significantly; patients were at least ten times more likely to die from variolation than from vaccination. But undeniably, a defining element of the intervention lay in the idea of triggering an immune response by exposing a patient to a small quantity of infected material. That idea had emerged elsewhere, not in the fertile mind of the country doctor, musing on the strange immunity of the milkmaids, but rather in the minds of pre-Enlightenment healers in China and India hundreds of years before. The fact that Jenner was able to modify the practice of variolation to utilize cowpox, not smallpox, was itself dependent on the diffusion of variolation through the British medical establishment. Rewind the tape of history and change one variable—Mary Montagu remains in London instead of moving to Constantinople—and it is entirely conceivable that variolation takes far longer to take hold as a medical practice in England.

Alternate histories are pure speculation, of course, but indulging in them forces us to think about the prime movers of change in society, and the importance of transmission vectors in making meaningful change in the world. Ideas are like viruses. For an idea to transform a society, the institutions and agents who transmit the idea are in many ways just as critical as the original minds that conceived the idea. Keep Mary Montagu in London and one fact is clear: variolation would have been forced to follow another path into the mainstream of British medicine. Perhaps its spread from East to West was inevitable, and the idea would have "infected" the minds of British doctors within the same time frame had Montagu not made the trip to Istanbul. But the practice had been thriving for centuries around the world without making its way across the Channel; without Montagu, it's certainly plausible that it might have remained there for another fifty years, long enough to radically change the history of British medicine, and delay that first spike in life expectancy that emerged in the second half of the 1700s.

On the one hand, we have the satisfying narrative of the brilliant Edward Jenner, inventing vaccination on one day in 1796. On the other, we have a much more complicated story, where part of an idea emerges halfway around the world, migrates from culture to culture through word of mouth, until a perceptive and influential young woman takes note of it and imports it to her home country, where it slowly begins to take root, ultimately allowing a country doctor to make a key improvement on the technique after decades of using it on his own patients.

You can think of these two kinds of narratives as the difference between the "genius" narrative and the "network" narrative. In the genius narrative, the causal chain revolves around the minds of one or two key pioneers, who single-handedly discover the breakthrough idea. The genius narrative is ubiquitous in the shorthand of history textbooks, which collapse what were truly network stories into the genius structure: Thomas Edison invents the light bulb or Alexander Fleming discovers penicillin. The network narrative is more complicated, in part because there are often simultaneous discoveries of the idea or technology in question. Variations on incandescent light were invented more than a dozen times in the 1870s; even Jenner's vaccine appears to have been preceded by a similar cowpox-based inoculation performed in 1774 by another rural English doctor, Benjamin Jesty. But the network narrative is more complicated as well because it emphasizes roles beyond that of the original discoverer in making a new idea valuable to the general public. An idea on its own is insufficiently powerful to transform society. Many great ideas die out before they can have a wider effect because they lack other key figures in the network: figures that amplify or advocate or circulate or fund the original breakthrough. Gregor Mendel, famously, had one of the great ideas of the nineteenth century, breeding his pea pods in his Moravian monastery. But because he was not connected to a wider network, the theory of genetics did not have any meaningful effect on the world for forty years.

"Actual examples of the lone genius phenomenon, in which an investigator single-handedly resolves a large problem, are few and far between," Cary Gross and Kent Sepkowitz write in a paper analyzing the network of innovations behind the smallpox vaccine. "Much more commonly, developments represent the culmination of decades, if not centuries of work, conducted by hundreds of persons, complete with false starts, wild claims, and bitter rivalries. The breakthrough is really the latest in a series of small incremental advances, perhaps the one that has finally reached clinical relevance. Yet once a breakthrough is proclaimed, and the attendant hero identified, the work of the many others falls into distant shadow, far away from the adoring view of the public."[7] The emphasis on a sudden breakthrough is not just a matter of historical inaccuracy; it distorts our priorities and our funding strategies in trying to encourage the next generation of innovations. "Disease-specific interest groups have had great success swaying public opinion, and research dollars, in their favor," Gross and Sepkowitz argue. "The public is enamored with the idea of the 'breakthrough'; a search for this word in the Nexus database yielded 1096 media citations over the past two years. A climate of unrealistic expectations by patients and the general public alike has developed. As such, research that does not overtly go for the 'home run' may be hampered and even endangered."[8]

The emphasis on networks is not just a matter of there being more characters on the stage. There is a qualitative difference as well. As you track the history of our doubled life expectancy, you begin to see certain *roles* in the network appear again and again. Mary Montagu performed two roles in the collaborative network that ultimately gave rise to vaccination, roles that are almost always performed in some fashion when new ideas take root in society. First, she was a *connector*, importing an idea from another domain, allowing the idea to cross both intellectual and geographic borders. And at the same time she was an *amplifier*, spreading word about the procedure through her writing and her influence among the British peerage and the royal family.

Interestingly, a similar connective pattern occurred with variolation in the American colonies, right around the same period, only with a different geography. Inoculation first arrived in New England via slaves who had a long history employing the procedure in their African homeland. Within a few years of Montagu's fateful visit to Turkey, a slave named Onesimus, believed to be of Sudanese descent, informed his master that he was not vulnerable to the smallpox. "People take juice of Smallpox, cut skin, and put in a drop,"[9] he explained. The master happened to be Cotton Mather, the influential Puritan preacher. Despite his belief in witches and devils—prominently on display during the Salem Witch trials—Mather had a meaningful interest in scientific investigations. Onesimus's account of his inoculation back in Sudan ultimately turned Mather into a firm believer in the power of variolation, even as some of his peers in the religious community objected to the practice. (Its 2 percent mortality rate was seen as violating the Sixth Commandment: Thou shall not kill.) Mather would go on to play a key role advocating for variolation among the growing colonies of New England, writing sermons and pamphlets, and proselytizing for the practice among the medical community in Boston. Onesimus served as the connector in this American version of the network narrative, importing a new idea from one culture to another, thanks to the brutal displacements of the slave trade. Cotton Mather took that idea and amplified it, using the power of the pulpit and the printing press.

For all their differences, Mary Montagu, Onesimus, and Cotton Mather had one notable quality in common: none of them were members of the medical profession. And yet they each had a significant impact on the adoption of variolation, thanks to their roles as connectors and amplifiers. This, too, turns out to be a common theme in the history of extending life: scientists and physicians are only part of the network that drives meaningful change. Without activists and reformers and evangelists, many life-saving ideas would have languished in research labs or been resisted by the general public. We have an understandable tendency to attribute the great escape exclusively to the triumph of enlightenment science. Once the great minds of Western culture began applying the scientific method to the problem of disease and mortality, we assume, the extension of life was an inevitable outcome. But the history of vaccination reminds us that this story is incomplete, not just because variolation itself emerged outside of the West. The triumph of vaccination was a matter of persuasion as much as it was empiricism. Important breakthroughs in health don't just have to be discovered; they also have to be argued for, championed, defended.

BECAUSE THEY WERE medical interventions that exposed otherwise healthy people to dangerous viruses, variolation and vaccination were particularly dependent on the support of influential early adopters like Mary Montagu. But the most remarkable advocate for the smallpox vaccine was an American who also lacked any medical background. In the early months of 1800, four years after Jenner's milkmaid experiment, a Harvard Medical School professor named Benjamin Waterhouse received a sample of smallpox vaccine that had been sent to him from a doctor across the Atlantic in Bath. Waterhouse had already published an essay on the new technique and was so confident in its efficacy that he inoculated his own family and then exposed some of them to smallpox patients to prove that the experiment had been a success. Yet Waterhouse sought a

larger platform for this medical breakthrough. And so he sent a letter to a well-connected amateur scientist in Virginia, enclosing his
essay "Prospect Of Exterminating The Small Pox."

The Virginian wrote back an enthusiastic note, and the two men
began a long-distance collaboration that would play a pivotal role
in bringing vaccination to mainstream American medicine. Three
times Waterhouse sent "vaccine matter" through the postal service,
but the Virginian reported that each time tests showed that the
vaccine had not survived the trip, likely due to the heat killing off
the live viruses. He proposed to Waterhouse an ingenious packaging design to preserve the vaccine: "Put the matter into a phial of
the smallest size, well corked and immersed in a larger one filled
with water and well corked," he wrote. "It would be effectually
preserved against the air, and I doubt whether the water would
permit so great a degree of heat to penetrate to the inner phial as
does when it is in the open air. It would get cool every night and
shaded every day under the cover of the stage, it might perhaps suc-
ceed."[10] The design worked, and by November 1801, the Virginian
was able to report in a letter that he had "inoculated about 70 or 80
of my own family, my sons in law about as many of theirs, and including our neighbors who wished to avail themselves of the opportunity. Our whole experiment extended to about 200."[11] He
took careful medical notes on the physical reaction to the vaccine,
which he dutifully sent back to Waterhouse:

> As far as my observation went, the most premature cases
> presented a pellucid liquor the sixth day, which continued
> in that form the sixth, seventh, and eighth days, when it
> began to thicken, appear yellowish, and to be environed
> with inflammation. The most tardy cases offered matter on

the eighth day, which continued thin and limpid the eighth, ninth, and tenth days.[12]

In the subsequent months he exposed a number of the vaccinated group to the smallpox virus and confirmed that all of them had developed immunity to it. While the experiments lacked the statistical sophistication of modern drug trials, they nonetheless marked a crucial leap forward in the adoption of vaccines: just five years after Jenner's breakthrough, hundreds of people were being successfully vaccinated across the Atlantic, with empirical evidence documenting the success of the trial. Given the general quackery of most medical science during this period, the vaccine trials would have been an astonishing achievement for a full-time doctor, but the Virginian was only moonlighting as a health professional. His day job, as it happens, was president of the United States, and his name, of course, was Thomas Jefferson.

As mind-boggling as it is to contemplate a sitting president conducting experimental drug trials in his spare time, there is something appropriate in a politician trained as a lawyer playing such a key role in the adoption of vaccination in the United States. In many respects, the story of vaccination's spread from a small vanguard of pioneers like Jefferson to mass adoption is a story of legal triumphs, not medical ones. The laws that mandated vaccination were milestones in the history of governance in that many of these laws marked the first time the state had exercised its power over individual health decisions. A decade or so after Jefferson's pioneering experiments, in 1813, Congress passed the Vaccine Act, with the aim to "furnish . . . genuine vaccine matter to any citizen of the United States."[13] In England, the Vaccination Act of 1853 required all children under three years of age to be administered a smallpox vaccine. (A series of subsequent acts over the following decades made the laws even more stringent.) Germany made vaccination compulsory in 1874.

The vaccination laws were written by elected officials, but the public support for them was often generated by advocates who were neither politicians nor public health officials. In many respects, the Mary Montagu and Cotton Mather of nineteenth-century vaccination was none other than Charles Dickens, whose classic novel *Bleak House* featured a critical plot twist that involves an unnamed disease that is clearly smallpox. Dickens published dozens of provaccination essays—many of them written by him—in his popular weekly magazine, *Household Words.* He was an impassioned advocate for compulsory vaccination, and frequently lionized Edward Jenner as one of the great heroes of modern life. "Few thoughts have given more material benefit to man," Dickens wrote in 1857, "than that which arouse in Dr. Jenner's mind, when it occurred to him that by putting intention in the place of accident, the benefit of exemption from small-pox might be extended."[14]

The vehemence of Dickens's support for mandatory vaccination was itself precipitated by the rise of a Victorian anti-vax movement, one that shares many of the same values of today's dissenters. Starting in the middle of the 1800s, a wave of pamphlets, books, satirical cartoons, court battles, loose alliances, and formal organizations arose in response to the perceived encroachment of mandatory vaccination. There was the Anti-Vaccination Society of America, the New England Anti-Compulsory Vaccination League and the Anti-Vaccination League of New York City. In England, the Anti-Compulsory Vaccination League was created in 1867, declaring that "Parliament, instead of guarding the liberty of the subject, has invaded this liberty by rendering good health a crime, punishable by fine or imprisonment, inflicted on dutiful parents."[15] The movement's leaders included some formidable intellectual figures, including Hebert Spencer and Alfred Russel Wallace, the latter of whom had famously developed the theory of natural selection independently in the 1850s. Wallace wrote several works late in his life attacking the science of vaccination, with such titles as *Vaccination Proved Useless and Dangerous* or *Vaccination a Delusion: Its Penal Enforcement A Crime.* It is odd to think that the codiscoverer of evolution might also have been the Jenny McCarthy of his age, but Wallace's tracts did attempt to mount an empirical case against vaccination based on public health data. In the long run, his work inspired the collection of better data sets in the early twentieth century, which ultimately made a convincing case for the effectiveness of the practice.

The anti-vaccination movement lay at the convergence point between three distinct currents. First, there were various forms of spiritualism, homeopathy, and "natural healing" that loomed so large in late-Victorian society. (Wallace had been a convert to spiritualism in the 1860s.) Another group opposed vaccination because it distracted the health authorities from what they considered to be the prime culprit in the spread of disease: unsanitary living conditions. (These were in many cases the descendants of the miasma theorists who had resisted the waterborne theory of cholera in the 1850s.) And then there were the political opponents, including Spencer, who saw in mandatory vaccination the ultimate encroachment of the state over individual liberty. University College professor F. W. Newman was often quoted making that case: "Against the body of a healthy man Parliament has no right of assault, whatever under pretense of the Public Health; nor any the more against the body of a healthy infant. To forbid perfect health is a tyrannical wickedness, just as much as to forbid chastity or sobriety. No law-giver can have the right. The law is an unendurable usurpation, and creates the right of resistance."[16]

The ideological resistance to vaccines has its roots in one of the intervention's unique qualities: vaccines were explicitly touted as a medicine for people who were not yet sick. Administering them was a purely preventative act, one of the very first that had the weight of science behind it. Delivering them to small children in perfect health seemed, intuitively, like a grotesque overreach, an act of "tyrannical wickedness." But that intervention had statistics on its side. If you received the vaccine as a child, you were much more likely to live long enough to have children yourself. In the long run, those odds won out over the protests of the anti-vaxxers.

The British protesters did manage to secure a clause in an 1898 act that allowed parents to receive a "certificate of exemption" if they claimed that vaccination went against their beliefs. The law marked the first time the concept of "conscientious objection" entered English law—a concept and phrase that would play an important role in the military conflicts of the twentieth century. Similar exemption clauses have become flashpoints in the recent controversies over anti-vax movements, with a number of local governments in the United States revoking exemptions after new outbreaks of measles—long considered eradicated in the United States—began to appear in communities with high proportions of anti-vax families. The difference, of course, between the nineteenth-century dissenters and their twenty-first-century descendants is the extraordinary global triumph of vaccination that took place in the century between them. The Victorian protesters had only the smallpox vaccine to consider, and limited statistical tools at their disposal to gauge its efficiency. The modern anti-vaxxer has a far more impressive track record to willfully ignore, both in terms of the ranges of diseases that vaccines now combat—diphtheria, typhoid, polio, and so on—but also the empirical evidence of the life-saving properties of these interventions. The best estimates hold that roughly a billion lives were saved thanks to the invention and mass adoption of vaccination over the past two centuries since Jenner's initial experiment. That extraordinary success was the product of medical science, to be sure, but also activists and public intellectuals and legal reformers. In many ways, mass vaccination was closer to modern breakthroughs like organized labor and universal suffrage: an idea that required social movements and acts of persuasion and new kinds of public institutions to take root.

ONE OF THOSE INSTITUTIONS dates back to a conference organized in Paris in 1851. Compared to the grand scale of most industry conventions today, the gathering was a modest affair comprising a physician and diplomat from each of twelve European nations The meeting became known as the International Sanitary Conference, and it marked one of the first times in history that a group of experts from a wide range of countries gathered to discuss ways to collaborate on public health. The 1851 conference focused on standardized quarantine procedures to limit the spread of cholera, but subsequent conferences widened their focus to include the sharing of emerging therapeutic techniques, epidemiological data, and scientific research on disease. The conferences ultimately led to the formation of the Office International d'Hygiene Publique (the International Office of Public Health [IOPH]) in Paris in 1907, one of the first truly international organizations ever created. After the founding of the United Nations in 1945, the IOPH was replaced by a new entity, formed under the UN umbrella: the World Health Organization, or WHO.

There is an unfortunate tendency, in a culture so obsessed with the creative destruction of technology start-ups, to assume that institutions are the enemy of innovation. If we want new ideas and progress and breakthrough technologies, the story goes, we need agile free agents who will move fast and break things, not ponderous, bureaucratic institutions. But viewed on a truly global scale, it is hard to find an entity that has done more to improve the lives of *Homo sapiens* over the past seventy years than the World Health Organization. And out of all the WHO's achievements over that period, one stands head and shoulders above the rest: the eradication of smallpox.

After thousands of years of conflict and cohabitation with humans, the naturally occurring variola major virus infected its last human being in October 1975, when the telltale pustules erupted on the skin of a three-year-old Bangladeshi girl named Rahima Banu Begum. Begum lived on Bhola Island, on the southern coast of Bangladesh, at the mouth of the Meghna River. WHO officials were notified of the case and sent a team to treat the young girl, and to vaccinate all the individuals on the island who had come into contact with her. She survived her encounter with the disease, and the vaccinations on Bhola Island kept the virus from replicating in another host. Four years later, on December 9, 1979, after an extensive global search for other outbreaks, a commission of scientists signed a document proclaiming that smallpox had been eradicated. In May of the following year, the World Health Assembly officially endorsed the WHO findings. Their proclamation declared that "the world and all its peoples have won freedom from smallpox,"

Rahima Banu Begum In her
mother's arms, 1975
*(Smith Collection, Gado / Alamy
Stock Photo)*

and paid tribute to the "collective action of all nations [that] have freed mankind of this ancient scourge."[17] It was a truly epic achievement, one that required a mix of visionary thinking and on-the-ground fieldwork spanning dozens of different countries. And yet the popular awareness of smallpox eradication pales beside that of achievements like the moon landing, despite the fact that eliminating this ancient scourge had a far more meaningful impact on human life than anything that came out of the space race. Just think of how many films and television series have celebrated the heroic, one-giant-leap-for-mankind daring of astronauts, and how few have chronicled the far more urgent—but equally daring—battle against lethal microbes.

The comparison between smallpox eradication and the space race is intriguing for another reason: because in many ways the battle against variola major was a triumph of global collaboration rather than competition, despite the fact that it took place during the Cold War. One of the early seeds of the project was planted in a 1958 speech at a gathering of the WHO in Minneapolis, delivered by Dr. Victor Zhdanov, deputy minister of health of the Soviet Union, calling for all the partner nations to commit to the then-audacious goal of eradicating smallpox. Zhdanov began his talk by quoting a letter Thomas Jefferson had written to Edward Jenner in 1806, which predicted that Jenner's smallpox vaccine would ensure that "future nations will know by history only that the loathsome small-pox has existed." In the subsequent two decades—through the downing of Francis Gary Powers's spy plane and the Cuban missile crisis and the Vietnam war—the United States and the USSR would somehow find a way to work productively together on smallpox eradication, a reminder that global cooperation on crucial issues in human health is possible even in times of intense political disagreement.

In another letter, composed during his early trials of the small-pox vaccine, Jefferson wrote to Benjamin Waterhouse: "It will be a great service indeed rendered to humanity to take off the catalogue of its evils so great a one as the small pox. I know of no discovery in medicine equally valuable."[18] Jefferson, as usual, was thinking long-term. The fight against variola major was progressing on a patient-by-patient basis when Jefferson wrote those words. The total number of people vaccinated in the world was in the thousands, maybe fewer. Removing smallpox from humanity's "catalogue of its evils" on a global scale was barely imaginable in 1801. Certainly, it was a technical impossibility. Science had progressed to the point where small-pox immunity could be triggered in one individual, with minimal risk. But eradicating smallpox as a disease across the entire world? We simply didn't have the tools to make it happen.

Thinking about the gap between Jefferson's early dream of removing smallpox from the "catalogue of evils" and the reality of eradication allows us to understand more clearly the forces that drive momentous change in the world. What did we have at our disposal in the 1970s that Jefferson and Waterhouse and Jenner didn't have back in 1801? What moved smallpox eradication from an idle fantasy to the realm of possibility?

One key factor was the institution of the WHO itself. The eradication project began in earnest with a proposal for the elimination of smallpox in West Africa, written by D. A. Henderson, then the head of disease surveillance at the Centers for Disease Control in Atlanta. The proposal caught the eye of the White House, and in 1965 Henderson was asked to move to Geneva to oversee a more ambitious program of global eradication for the WHO. Even Henderson himself thought the program would likely end in failure, given the audacity of the goal. But he ultimately took the assignment and would oversee the program until the eradication certification was signed in 1979. During the decade of active surveillance and vaccination, the WHO worked in concert with seventy-three different nations, and employed hundreds of thousands of health workers who oversaw vaccinations in the more than two dozen countries still suffering from variola major outbreaks. The idea of an international body that could organize the activity of so many people over such a vast geography, and over so many separate jurisdictions, would have been unthinkable at the dawn of the nineteenth century. Global eradication was as dependent on the invention of an institution like the WHO as it was on the invention of the vaccine itself.

The eradication program also depended on a relatively new insight from the domain of microbiology—a science that did not exist in any serious sense during Jefferson's age. (The variola major virus would not be identified via microscope for another century or so.) But by the time D. A. Henderson first began formulating his eradication plans, virologists had come to believe that the smallpox virus could only survive and replicate inside human beings. The virus, in the technical language, had no remaining natural reservoir in other species. Many viruses that cause disease in humans can also infect animals—think of Jenner's cowpox. But smallpox had lost the ability to survive outside of human bodies; even our close relatives among the primates are immune. This knowledge gave the eradicators a critical advantage over the virus. A traditional infectious agent under attack by a mass vaccination effort could take shelter in another host species—rodents, say, or birds—that would be impossible for WHO field-workers to survey and, if infected, eliminate. But because smallpox had abandoned whatever original host had brought it to humans, the virus was uniquely vulnerable to Henderson's campaign. If you could drive variola major out of the human population, you could truly remove it from the catalogue of evils for good.

Technical innovations also played a crucial role in the eradication projects. The invention of the bifurcated needle allowed the WHO field-workers to use what was called a multiple puncture vaccination technique. It was both much easier to perform and required a quarter of the amount of vaccine as earlier techniques, essential attributes for an organization attempting to vaccinate hundreds of thousands of people around the world. Another crucial asset—unavailable to Jefferson and Waterhouse—was a heat-stable vaccine, developed in the 1950s, that could be stored for thirty days unrefrigerated, an enormous advantage in distributing vaccines to small villages that often lacked refrigeration and electricity.

The last innovation revolved around the method of mass vaccination itself. In December 1966, not long after D. A. Henderson had taken the helm of the smallpox eradication program at the WHO, an epidemiologist named William Foege, working for a CDC program, found himself battling an outbreak in the Liberian village of Ovirpua. The typical response to such an outbreak would be to vaccinate every single member of the village (as well as nearby villages). But the CDC program was new and sufficient supplies of vaccine had not yet been delivered. Given the limited resources, Foege was forced to improvise a solution that could do more with less. As he later described it in his memoirs, Foege and his colleagues asked themselves, "If we were smallpox viruses bent on immortality, what would we do to extend our family tree? The answer, of course, was to find the nearest susceptible person in which to continue reproduction. Our task, then, was not to vaccinate everyone within a certain range but rather to identify and protect the nearest susceptible people before the virus could reach them."[19] Instead of dumping a massive amount of vaccine on an entire region, Foege decide to create what he called a ring of vaccinations that would surround the infected villagers. It was a targeted strike, designed to build a firewall of immunity around the outbreak. To Foege's surprise, it worked. Within a matter of days, the outbreak had ended. Foege's "ring vaccination" technique ultimately became the basis of the WHO global eradication project. When Rahima Banu Begum contracted smallpox on Bhola Island in 1975, it was a firewall of vaccination around her that ended the scourge of variola major once and for all.

ONE OF THE REASONS Foege's ring vaccination approach wasn't available to Jenner and Waterhouse back in the early 1800s is that it relied on a specific way of thinking about illness. The difference was, literally, a matter of perspective: most attempts at combating illness in Jenner's age were centered on the human body itself, with its mysterious machinery—veins and lungs and muscle and all the rest. But the ring model looked at the problem from a different vantage point. You could attack the disease by looking at its distribution geographically, as it spread from person to person, and from community to community. Jenner couldn't think that way because the science of epidemiology didn't exist in a coherent form during his time. People understood that diseases clustered in meaningful patterns, and they had tried to map those outbreaks in a crude fashion. But they had not yet turned those maps into a weapon that could be used against the infectious agent itself. On one level, Foege's breakthrough idea of ring vaccination was a classic case of necessity being the mother of invention: with limited supplies of vaccines, he was compelled to seek out a different solution. But the idea also came to him because he was trained in a discipline that had more than a hundred years of practical and experimental thinking behind it. His mind drifted to the bird's-eye view because he was trained as an epidemiologist. That science turns out to be the last key ingredient that separated Jenner and the eradicators. As is so often the case in the history of progress, a shift in the way we perceived the problem turned out to offer a kind of solution in itself.

And not just for the smallpox eradicators. Variolation might have propelled the British aristocrats to longer lives in the late 1700s, but it was the data revolution of epidemiology—and not vaccination—that gave us the first sustained increase in life expectancy that mattered to the masses.

3

VITAL STATISTICS

DATA AND EPIDEMIOLOGY

The River Lea originates in the suburbs north of London, winding its way southward until it reaches the city's East End, where it empties into the Thames near Greenwich and the Isle of Dogs. In the early 1700s, the river was connected to a network of canals that supported the growing dockyards and industrial plants in the area. By the next century, the Lea had become one of the most polluted waterways in all of Britain, deployed to flush out what used to be called the city's stink industries.

In June 1866, a laborer named Hedges was living with his wife on the edge of the Lea, in a neighborhood called Bromley-by-Bow. Almost nothing is known today about Hedges and his wife other than the sad facts of their demise: On June 27 of that year, both of them died of cholera.

The deaths were not in themselves notable. Cholera had haunted London since its arrival in 1832, with waves of epidemics that could kill thousands in a matter of weeks. While the disease was on the decline in recent years, a handful of cholera deaths had been reported in the preceding weeks, and it was not unheard of for two people sharing a home to die of the disease on the same day.

But the deaths of Mr. and Mrs. Hedges turned out to be the start of a much bigger outbreak. Within a few weeks, the working-class neighborhoods surrounding the Lea were suffering one of the worst cholera epidemics in London's history. The newspapers delivered the same sort of morbid accounting that has obsessed us all in the age of the coronavirus: the terrifying upward trajectory of runaway growth. Twenty cholera deaths were reported in the East End the week ending July 14. The following week's tally was 308. By August, the weekly death toll had reached almost a thousand.[1] London had not experienced a major outbreak of cholera for twelve years. But by the second week of August, the evidence was unmistakable: the city was under siege.

Just as we have seen in the age of COVID-19, the first line of defense against the outbreak was data. Londoners were able to track the march of cholera across the East End in close to real time, thanks primarily to the work of one man: a doctor and statistician named William Farr. For most of the Victorian era, Farr oversaw the collection of public health statistics in England and Wales. You could say without exaggeration that the news environment that emerged during the height of the COVID pandemic was one that William Farr invented: a world where the latest numbers tracking the spread of a virus—how many intubations today? What's the growth rate in hospitalizations?—became the single most important data stream available, rendering the old metrics of stock tickers or political polls mere afterthoughts.

Farr was among the first to think systematically about how data on outbreaks, their distribution in space and over time, could be used to curb them as they unfolded—and to minimize future ones. The field he helped invent has come to be called epidemiology, but in its infancy it was known by another name: vital statistics. (*Vital* as in *vita*, Latin for *life*.) The innovations in this field do not look like our traditional model of medical breakthroughs. They are not packaged in the form of miracle drugs or new imaging technologies. At their core, they are simply new ways of counting, new ways of discerning patterns.

WHEN THE EAST END outbreak first became apparent to the health authorities—and to the terrified residents of the neighborhoods under assault—it seemed to be a continuation of a broader mortality trend that had afflicted industrial towns and neighborhoods for most of the century. While the elites of England had experienced an unprecedented increase in life expectancy after 1750, largely fueled by variolation and vaccination, equally significant during that period was the utter lack of health progress among the less fortunate classes of society. Variolation and vaccination had spread through the rural poor and the industrial working classes during that period as well. And yet the mortality rates in those groups stayed constant over that period, even crept downward in places. If you were wealthy, your life expectancy had been extended by almost thirty years over that stretch. If you were poor, you were no better off than you would have been in John Graunt's era.

The mortality trends in the United States during the first half of the nineteenth century were even more stark. Despite the widespread adoption of vaccination, overall life expectancy in the United States declined by *thirteen* years between 1800 and 1850. The twin revolutions of Enlightenment science and industrialization had transformed both England and her former colonies across the Atlantic, creating new economic and political systems, driving immense technological change: factories, railroads, telegraphs. Yet where life expectancy was concerned, the world's most technologically advanced societies appeared to be going in reverse. Meaningful improvements to overall mortality would not arrive until the last decades of the nineteenth century, previewing the dramatic takeoff that would transform life expectancy on a global scale in the next century.

This pattern poses two fascinating questions: Why did mortality rates in such advanced societies—societies that should have been enjoying the benefits of Enlightenment reason—go backward for half a century? And when the great escape finally began in earnest, lifting overall life expectancy in those closing decades, what developments drove that change? The answer to both those questions would turn out to be at play in the East End outbreak of 1866.

The first question—Why were the industrial poor dying?—was one that was actively pondered and investigated at the time. In a sense, the modern science of epidemiology can be said to have originated as an attempt to solve this mystery. Arguably the most influential detective investigating the case was William Farr. Born in 1807 into a rural family of little means, Farr was a precocious learner who attracted the support of a few wealthier patrons and mentors as a teenager, apprenticing with a local surgeon before studying medicine in Paris and at University College in London. By his mid-twenties, Farr had established a medical practice in London. But his true passion was for vital statistics: the analysis of births and deaths in a large population. In many ways, Farr's long and illustrious career marks the culmination of the idea that John Graunt had first sketched in his *Natural and Political Observations*: that perceiving the macro patterns in mortality could become a life-saving tool itself, as effective as any traditional medical intervention.

In his passion for statistics and for social reform, Farr was very much a man of his time. A number of "statistical societies" had been formed in British cities during the 1830s; Farr himself was an early member of the London Statistical Society. The use of data to understand patterns of life and death had been almost exclusively a commercial interest during the eighteenth century, a science developed largely for the mercenary aims of the insurance companies. But Farr and some of his peers saw the potential of vital statistics as a tool for social reform, a means of diagnosing the ills of society and shining light on its inequalities.

After publishing a few papers analyzing medical data in *The Lancet*, Farr was hired in 1837 as a "Compiler of Abstracts" at the General Register Office (GRO), a newly created government body tasked with tracking births and deaths in England and Wales. At Farr's encouragement, the GRO began recording a much wider range of data in its mortality reports, including cause of death, occupation, and age. In a letter appended to the first GRO report, Farr laid out his ambition for the office. "Diseases are more easily prevented than cured," he wrote, "and the first step to their prevention is the discovery of their exciting causes. The Registry will show the agency of these causes by numerical facts and measure the . . . influence of civilization, occupation, locality, seasons and other physical agencies whether in generating diseases and inducing death or in improving the public health."[2] Farr helped create a systematic classificatory scheme for causes of death, a great improvement on the erratic scheme—"Cut of the stone, Lunatick, Suddenly"—that Graunt had employed. Farr also helped to establish the first proper census of the English population, conducted in 1841, giving the GRO another crucial data set that it could use to understand the overall state of the nation.

As a compiler of abstracts, Farr was responsible for taking the raw data recorded by the GRO and making it meaningful: discovering interesting trends in the numbers, comparing health outcomes for different subgroups in the population, inventing new forms of visualization. Collecting and publishing data was not merely a matter of reporting the facts, but instead a more subtle, exploratory art: testing and challenging hypotheses, building explanatory models. As Farr wrote in an essay published the year he joined the GRO, "Facts, however numerous, do not constitute a science. Like innumerable grains of sand on the sea shore, single facts appear isolated, useless, shapeless; it is only when compared, when arranged in their natural relations, when crystallized by the intellect, that they constitute the eternal truths of science."[3]

The specific arrangement of facts that Farr relied on more than any other was a descendant of the original "life table" that John Graunt had built in his 1662 pamphlet: a chart that breaks down the mortality rates of a given population by age group. Comparing the life tables of distinct communities gave a clear picture of the differences in health outcomes between them. Edwin Chadwick, the pioneering public health reformer of the period, had proposed a simpler measure of community health: the mean age of death. But as Farr pointed out multiple times, reducing the patterns of deaths in a community to a single number could be misleading, particularly when comparing that data point to the average age of death in another community. Given the high infant and child mortality rates of the period, a town that happened to be going through a period of high fertility with more children being born would paradoxically have a lower mean age of death than a town with a higher proportion of adults—even if the former community happened to be healthier. (Despite the health of the community, a significant number of the children would still die before reaching adulthood, which would then drag down the average age at death.) A life table allowed you to see in a glance what was really happening in a given population: both the big picture and the age-by-age breakdown.

Perhaps because of his own personal itinerary—growing up in the agricultural region of Shropshire, now living in the largest city on the planet—Farr decided to devote one of his first studies to the differences in health outcomes between the country and the city. In the *First Annual Report* of the GRO, published in 1837, Farr authored a section called "Diseases of Town and the Open Country." In it, he drew upon some rough data sets he had assembled for London and for some rural districts in the southwest of England. He continued tinkering with the analysis in subsequent annual reports, culminating in a groundbreaking study featured in the *Fifth Annual Report*, released in 1843. Farr's study marked a milestone in the emerging science of epidemiology; relying on a pool of data that Farr himself had made possible through his work with the General Registry and the census, the study also showcased Farr's imaginative use of data visualization techniques.

The 1843 study analyzed three separate communities: metropolitan London, industrial Liverpool, and rural Surrey. It was, in effect, a tale of two cities—and one countryside. Viewed as a triptych, the illustrations conveyed a clear message: density was destiny.

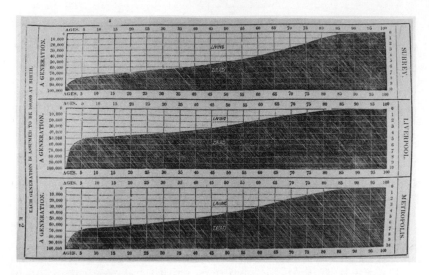

William Farr's life tables

In Surrey, the increase of mortality after birth is a gentle slope upward, a dune rising out of the waterline. The spike in Liverpool, by comparison, looks more like the cliffs of Dover. That steep ascent condensed thousands of individual tragedies into one vivid and scandalous image: in industrial Liverpool, more than half of all children born were dead before their fifteenth birthday.

The mean age of death—keeping in mind its limitations as a data point—was just as shocking: the countryfolk were enjoying life expectancies close to fifty, a significant improvement over the long ceiling of the mid-thirties. The national average was forty-one. London, for all its grandeur and wealth, had retreated to the equilibrium of thirty-five, exactly where it was when Graunt first tried to measure it. But Liverpool—a city that had undergone staggering explosions in population density, thanks to industrialization—was the true shocker. The average Liverpudlian died at the age of twenty-five, one of the lowest life expectancies ever recorded in that large a human population.

The infographics of the *Fifth Annual Report* composed the first empirical argument for an idea that had seemed apparent to many people anecdotally: cities were killing people at an alarming, escalating rate. And they were particularly ruthless with young children. "The children of the idolatrous tribe who passed then through the fire to Moloch scarcely incurred more danger than is incurred by children born in several districts of our large cities," Farr warned. Echoing Jefferson's catalogue of evils language, he went on to write: "A strict investigation of all the circumstances of the children's lives might lead to important discoveries, and may suggest remedies for evils of which it is difficult to exaggerate the magnitude."[4]

This was the answer to the first question. Why were the most advanced nations in the world seeing their life expectancies decrease? How could any economy that was creating more wealth than any other place on earth produce such devastating health outcomes? The answer that Farr proposed with epidemiological data was similar to the one Marx and Engels were forming at the same time using political science. the mortality rates were plunging because the defining characteristic of being "advanced" at that moment in history was industrialization, and industrialization seems to come with an unusually high body count in its initial decades, wherever it happens to arrive. The twentieth century would go on to show the same trends happening around the world whenever people left their agrarian lifestyle and crowded into factories and urban slums, even in economies where communist planners were driving the shift to an industrial economy. Farr and his peers just happened to be seeing the pattern as it emerged for the first time.

Viewed from the perspective of the *longue durée*, the pattern Farr uncovered in his life tables of Surrey, London, and Liverpool contained two contrasting messages, one hopeful and one deeply troubling. Those rural populations in Surrey, with expected life climbing into the fifties, proved that human societies could break through the long ceiling of thirty-five years. And at the same time, those steep cliffs of Liverpool's childhood mortality rates made it clear that other kinds of societies could drop through the historic floor, plunging to rates only seen during the worst outbreaks of plague in England's past. The data told an incontrovertible story: industrial cities were killing people at an unprecedented rate. The great question of the age was whether that death toll—and all the other misery that accompanied it—was an inevitable by-product of factory towns and metropolitan density. Or was there a way of reversing the plunge?

From our vantage point today, the answer seems obvious: industrial cities were not inevitably doomed to be mass killers. Today many cities that sustain populations in the tens of millions enjoy some of the highest life expectancies—and lowest infant mortality rates—on the planet. But the answer was in fact already visible by the end of the nineteenth century. Starting in the 1860s, the industrial cities of England began to see a meaningful decline in mortality, a decline that for the first time was shared across the entire population, and not just clustered in rural communities or the aristocratic elite. That decline marked the true origin point of the great escape, a demographic transformation that would extend to the entire world in the next century. With hindsight, the East End cholera epidemic of 1866 turned out not to be a continuation of the dismal decades of industrial-age mass mortality. Instead, it turned out to be the beginning of the end. And the most important advance that drove that first sustained spike in life expectancy did not come out of medicine or health care. More than anything else, the first stirring of the great escape was a triumph of data.

In the 1843 report, Farr also turned his attention to another puzzling pattern in the data he had collected: what he called the laws of action of epidemics, now known to epidemiologists as Farr's laws. Analyzing a smallpox outbreak in Liverpool, Farr divided the mortality counts into ten separate periods. "The mortality increased up to the fourth registered period; the deaths in the first were 2,513; in the second, 3,289; in the third, 4,242; and it will be perceived at a glance that these numbers increased very nearly at the rate of 30 percent." But the rate of increase, he observed, "only rises to 6 percent in the next, where it remains stationary, like a projectile at the summit of the curve which it is destined to describe."[5] Farr's law was the first attempt to describe the rise and fall of contagious diseases mathematically. All the models that have shaped so much private angst and public scrutiny during the coronavirus pandemic—the Imperial College London models that steered Prime Minister Boris Johnson away from the initial strategy of herd immunity, the University of Washington COVID-19 projections that have heavily influenced the Trump White House—all these forecasts are descendants of the laws of action that Farr originally sketched in 1843. When we talk about flattening the curve, the curve in question was first drawn by William Farr.

IF YOU ASK most medical historians to pinpoint a crucial turning point in London's relationship to cholera—and the broader battle against urban mortality—they will not point you to late June 1866. The far more familiar milestone is September 8, 1854, the day that a local parish board in the London neighborhood of Soho removed the handle of a pump at 40 Broad Street in an attempt to stop one of the most devastating outbreaks of cholera in the history of the city. The handle had been removed at the urging of Dr. John Snow, who had been arguing for more than four years that cholera was a disease caused by contaminated water supplies, and not conveyed through polluted air, as the prevailing "miasma" theory contended. When cholera erupted on his doorstep in the last week of August, Snow recognized immediately that the intense concentration of the outbreak suggested that there might be a single "point source" that was causing people to become sick, and that some rapid investigative work determining the location of the dead—and their drinking habits—might reveal a specific source of contaminated water. Identifying that source might both end the outbreak, and finally convince the authorities that Snow's waterborne theory was correct. As part of his investigation, Snow famously constructed a map of the Broad Street outbreak, representing each death in the neighborhood with a little black bar placed at the residence associated with the deceased. The map deserves its pride of place as one of the most influential works of cartography in history, as important, in its own way, as the early maps that led the navigators around the world during the Columbian exchange. In those stacked black bars scattered across the street grid of Soho, Snow was trying to visualize something that was as remote from human perception as the coastline of the Americas was to Europeans before 1492: the transmission patterns of the microscopic agents that caused cholera in humans.

Snow had long suspected that the drinking water of London contained some kind of microorganism that triggered the violent diarrhea that killed cholera victims, and he spent hours in his home laboratory viewing samples of water from various sources through his microscope. But the lens-making technology of the age was not sufficiently advanced to allow him to see the bacterium—*Vibrio cholerae*—that we now know causes the disease. (It would be another three decades before the German microbiologist Robert Koch identified the bacterium.) But Snow recognized that there were other ways of seeing the agent. Instead of the microscope's zoom, he adopted the bird's-eye view, perceiving the agent indirectly, through the spatial distribution of the deaths it caused. Like Farr's life-table charts of Surrey and Liverpool, Snow built an empirical case using tools of data visualization. But Farr's life tables only suggested a general problem to be solved: something about dense urban living was killing people at an alarming rate. Snow's map, on the other hand, suggested a specific cause—and a specific remedy. People were dying because they were drinking water that had been contaminated, not because they were breathing in noxious fumes. If we wanted people to stop dying, we needed to clean up the water supply.

The removal of the Broad Street pump handle—and Snow's pioneering map—serves as a useful point of origin for the late Victorian revolution in public health for two reasons. While that revolution involved a number of different interventions, by far the most important one involved decontaminating public supplies of drinking water. And the story of the pump handle demonstrates that useful health interventions can be made without the benefit of actually understanding—or even being able to see—the biological mechanisms that cause epidemics.

But the story also makes for a good milestone thanks to its narrative power: a rogue medical detective, challenging the authorities in the midst of unbelievable terror, whose sleuthing and empirical method ends up transforming our understanding of disease and saving untold millions in the decades that follow. As it happens, I have personal connection to Snow's story, in that many years ago I wrote an entire book about the 1854 outbreak.[6] I had originally been drawn to this history precisely because it seemed to be one of those classic "lone genius" stories, with Snow in the leading role: one outsider's battle with the powers that be. That was largely how it had been told in most popular accounts up until that point.

But once I sat down to fully research the book, I quickly found that the lone genius model significantly distorted the causal forces that came together to remove the pump handle and overthrow the miasma theory more generally. It condensed a network down to a single individual. One member of that extended network undeniably was William Farr. Snow relied heavily on the data-collection techniques that Farr had pioneered in the preceding two decades and was very much in an intellectual dialogue with the statistician throughout his cholera investigations. Another member of the network was a local vicar named Henry Whitehead. Following in the footsteps of Cotton Mather's tenure as a variolation advocate, Whitehead got involved in the case running an amateur parallel investigation to Snow's: first trying to disprove the theory of the contaminated well, and then finding himself increasingly persuaded by the data. Eventually Whitehead became Snow's partner, and in fact it was Whitehead who discovered the "patient zero" of the case, a six-month-old girl—known only as Baby Lewis—who had come down with cholera at 40 Broad Street, contaminating the well water with her waste. (The well shared a decaying brick wall with the cesspool in the basement of 40 Broad.) Whitehead was also able to assemble additional data about deaths in the neighborhood—as well as track down locals who had fled Soho and died in the countryside—thanks to his deep roots in the community. A convincing argument can be made that without Whitehead's contributions, Snow's investigation into the Broad Street outbreak would not have convinced the authorities that his waterborne theory was correct, and the reigning miasma orthodoxy might have stayed in place for decades longer than it did. As is so often the case in meaningful social change, the revolution in our understanding of the relationship between water and disease required multiple actors with multiple skills: Farr's open-source data platform; Snow's epidemiological detective work and cartographic skills; Whitehead's social intelligence.

Knowing that cholera resulted from contaminated water supplies was only part of the solution. To actually do something about the disease, London had to rid its drinking water of the *Vibrio cholerae* bacterium: by separating the waste systems of the city from its water supplies. And that necessitated the building of one of the nineteenth century's greatest engineering achievements: the London sewer system.

Overseen by the brilliant and indefatigable Joseph Bazalgette, the project took the entirely haphazard network of drainage and waste pipes that had been accumulating for centuries beneath the city's streets, and replaced them with an organized system of sewer lines running eighty-two miles in total, using three-hundred-million bricks, including the massive interception lines that run along both banks of the Thames, keeping the city's waste from flowing downhill into the river. (The tourists strolling along the Victoria or Chelsea Embankments, gazing out at London's bustling skyline, are unwittingly enjoying a structure built specifically to keep the city's drinking water free of *Vibrio cholerae*.) Amazingly, the main lines of the project were functional after only six years of work.

The interesting footnote to this story is that William Farr remained a believer in the miasma theory for far longer than we might
have expected him to, given the data that he himself assembled and
his general appetite for testing and challenging his hypotheses.
Throughout his career, Farr clung to a strange bias against human
settlements at low elevations, a bias that had emerged in part out of
data he had assembled showing higher mortality rates near the banks
of the Thames. Farr believed that there were noxious fumes that
contaminated the air in the marshy boundaries between land and
water, and he produced a number of ingenious charts combining
mortality rates and topographic maps to prove the point. (Ultimately, the causal link between elevation and disease was shown to
lie, once again, in the drinking water: the farther you lived from the
Thames, the more likely you were to get your water from a less
contaminated source.) Farr's bias toward higher elevations ultimately
metastasized into a bizarre form of topographic racism, where the
highest achievements of civilization only emerge in cultures living
on higher ground. "The people bred on marshy coasts and low river
margins, where pestilence is generated, live sordidly, without liberty,
without poets, without virtue, without science," Farr wrote, in one
particularly shocking passage. "They neither invent nor practice the
arts; they possess neither hospitals, nor castles, nor habitations fit to
dwell in. . . . They are conquered and oppressed by successive trips
of the stronger races, and appear to be incapable of any form of society except that in which they are slaves."[7]

Farr's elevation theory of disease made little sense both medically and historically: one need only to think of Venice or the great
civilizations of the Nile Delta to recognize how wrong he was.
But despite the strange hold topography had over his interpreta-
tive models, he did ultimately become a believer in the waterborne

theory of cholera. Farr's conversion to Snow's theory would be put to the test in particularly dramatic fashion during the summer of 1866, just as Bazalgette's team was finishing work on the London sewers.

Now in his mid-sixties, Farr was still helping to oversee the production of the annual reports of the GRO, as well as the Weekly Returns of Births and Deaths that Snow had relied on in his Broad Street investigation. Scanning through the *Returns* in July 1866, Farr noticed the strange spike in cholera deaths in the East End. The disease had been close to dormant since the 1854 epidemic, and Bazalgette's sewers were largely operational, making the outbreak even more puzzling. A younger Farr might have turned his focus immediately to the topographic maps, calculating where the deaths lay in relation to sea level. But Farr in his mid-sixties was a different man, a reformed miasmatist. He had watched Snow build his case for the waterborne theory firsthand. With the death rate increasing week by week, Farr didn't bother running the numbers of elevation data; instead, he immediately began investigating the sources of drinking water in the neighborhood.

By the mid-1860s, a significant portion of even working-class communities were receiving their water through private companies that ran the pipes to specific addresses, much like the cable companies do today. Farr decided to sort the population that had died in the recent outbreak not by residence, but rather by the company that supplied their drinking water. The first rough pass of assembling the data revealed a clear pattern: an overwhelming number of cases were people who drank out of East London Waterworks Company pipes. Within days, Farr had placards posted throughout the east end, warning residents not to drink "any water which has not been previously boiled."

The investigation then turned to the East London company, which claimed its water had been effectively filtered at its new covered reservoirs. One of the lead investigators on the case had read a memoir of the 1854 Broad Street outbreak written by the vicar who had assisted John Snow in his surveys of the community. With Snow dead, the investigator thought his former partner might be helpful tracking down the cause of the East End outbreak. And that was how the Reverend Henry Whitehead found himself once again doing shoe-leather detective work on the streets of London, hunting down a hidden killer. By August they had uncovered the lines of contamination: one of the East London company's reservoirs had not been properly isolated from the nearby River Lea. Poring through the Weekly Returns from earlier in the summer, the investigators discovered the deaths of Mr. and Mrs. Hedges, who lived near the reservoir. An examination of their residence revealed that their water closet was expelling waste directly into the River Lea.

The story of Snow and the pump has been rightly considered a founding moment of modern epidemiology and public health, one of those moments in history where human beings flip a new kind of switch. But only some of the key pieces were in place during the Broad Street epidemic. In many ways, the 1866 outbreak should be considered just as important a milestone. In 1854, the state actors were only marginally important: Snow was an outsider, and most of the public authorities were still in the throes of miasma. Yes, Farr had built the mortality reports, but beyond that, the public sector figures were more hindrances than anything else. By 1866, however, the whole system had come together: Bazalgette had built his intercept lines; Farr had his data; the waterborne theory had become an accepted model among most of the public health decision makers. That integrated system was able to quickly detect a new outbreak, successfully contain it, and implement changes to the existing water delivery architecture that prevented future outbreaks from developing.

The success proved to be enduring. The East London crisis turned out to be the last outbreak of cholera ever recorded in London. *Vibrio cholerae* had arrived in 1832, after a long march across Europe. For a decade or two it threatened to become a killer on the order of smallpox or tuberculosis. And then it was gone. For London at least, cholera had been excised from the catalogue of evils, never to return.

FARR AND SNOW, in their different ways, made it clear that the deft use of vital statistics created new ways of seeing the realities of sickness and health in human populations. Farr's life tables exposed the inequalities in life expectancy that afflicted the urban centers; Snow's map revealed the waterborne nemesis that was causing cholera, even though the bacterium itself was not yet visible to science. At the very end of the nineteenth century, another data pioneer used maps and shoe-leather epidemiology to create a comparable breakthrough in our perception of human health: the polymath African-American public intellectual W. E. B. Du Bois.

Today, Du Bois is best known as a civil rights activist, as the founder of the NAACP and author of the seminal book on the African-American experience *The Souls of Black Folk*. But Du Bois's career began with a study of a black neighborhood in Philadelphia, published in 1899 in book form as *The Philadelphia Negro*. While the book has rightly come to be seen as a groundbreaking work of sociology, anticipating many of the techniques used by the Chicago

School in subsequent decades, it also deserves credit for helping to invent a new way of thinking about public health, a discipline sometimes called social epidemiology. Du Bois was the first to demonstrate a disheartening fact that has continued to haunt the United States in the age of COVID-19: African Americans were dying at higher rates than their white counterparts, and part of the explanation for that disparity lay in the way their lived environment was shaped by the oppressive forces of racism.

Du Bois was in his late twenties when he arrived in Philadelphia in 1896 for a one-year term as an "assistant in sociology." He had just completed his PhD at Harvard—the first African American to do so—and had spent a heady two years in Europe doing graduate work at the University of Berlin. In his early days as an undergraduate at Harvard, he had studied philosophy under luminaries like William James and George Santayana. But James had warned Du Bois that a career as a philosopher was challenging for any man "without independent means," and the political climate of the 1890s had increasingly compelled him to focus his prodigious intellect on what was then called "the Negro problem." A wave of sensationalist journalism and faux scholarship, with titles like "Race Traits and Tendencies of The American Negro," had addressed high rates of poverty and crime among African Americans, mostly by pointing to alleged deficiencies in the "Negro race" itself. Du Bois began to think that the embryonic tools of sociology might be a way of seeing the challenges of African-American communities through a scientific, data-driven lens free of the explicit prejudice that marred so much of the commentary on the "Negro problem."

In Philadelphia, a number of the city's well-to-do progressives—many of them belonging to the city's long-standing Quaker population—had been watching the rising crime and poverty in the city's Seventh Ward with an escalating sense of alarm. The Seventh Ward constituted a grid of eighteen blocks bordering the Schuylkill River. Like the Soho and East End neighborhoods that Farr and Snow analyzed, the neighborhood today is a prosperous mix of high-end restaurants and boutiques and renovated townhouses, but its nineteenth-century incarnation was a bleak landscape of urban degradation. Unlike those London neighborhoods, however, the crisis developing in the Seventh Ward had clear racial overtones: in the decades following the end of the Civil War, the neighborhood had become the largest African-American community in Philadelphia. "Because so many [African Americans] lived there," Du Bois's biographer David L. Lewis writes, "because many of them were so poor, because many had recently arrived from the South, because they were responsible for so much crime, and because they stood out by color and culture so conspicuously in the eyes of their white neighbors, the area was the bane of respectable Philadelphia, its population the very embodiment of 'the dangerous classes' troubling the sleep of the modernizing gentry."[8]

By 1895, it had become clear to the Philadelphia elite that the Seventh Ward—and other predominantly African-American neighborhoods in the city—were caught in a cycle of poverty and violence, what would eventually come to be called the crisis of the "inner cities" in the following century. As it happened, one of the white Philadelphians living near the Seventh Ward was the progressive philanthropist Susan Wharton, who had spent the previous ten years funding a range of charitable institutions with the aim of benefiting the city's African-American community. Wharton convinced the provost of the University of Pennsylvania that a "trained observer" was needed to make sense of the seventh ward's escalating problems—ideally, an observer who was African American himself.

Given his stellar academic record and his recent sociological work in Europe, Du Bois was the obvious man for the job. And so, in the summer of 1896, Du Bois and his wife moved into a one-room apartment at 700 Lombard Street on the eastern edge of the ward.

ON THE SURFACE, Du Bois's job description, as outlined by the UPenn faculty, had suggested an empirical mode of inquiry: "We want to know precisely how this class of people live; what occupations they follow; from what occupations they are excluded; how many of their children go to school; and to ascertain every fact which will throw light on this social problem."[9] But Du Bois went into the project well aware that his sponsors—for all their progressive ideals—still clung to racist beliefs about the core nature of the problem. Du Bois would later describe the tacit assumption as "Something is wrong with a race that is responsible for so much crime." And so the young scholar decided to combat that prejudice with a prodigious display of sociological detective work, an even more comprehensive and penetrating study of the neighborhood than the one John Snow had conducted in Soho forty years before. "The problem lay before me," he would later write. "I studied it personally and not just by proxy. I sent out no canvassers. I went myself. . . . I went through the Philadelphia libraries for data, gained access in many instances to private libraries of colored folk. . . . I mapped the district, classifying it by conditions." For months, Du Bois would leave 700 Lombard each morning, "accoutered with cane and gloves," and commence an eight-hour exploration of the seventh ward, knocking on doors, interviewing residents about their work lives and families, inspecting the conditions of their residences.

By the end of the survey, Du Bois had spent more than eight hundred hours documenting the conditions of the neighborhood, visiting more than two thousand households in just three months of research. Some of the data that he tabulated from the investigation would later be presented, like Snow had before, in the form of a map, each parcel in the ward color-coded to denote five classes of occupants: what Du Bois called "the vicious and criminal classes," "the poor," "the working people," "the middle classes," and the residences belonging to whites or commercial enterprises.

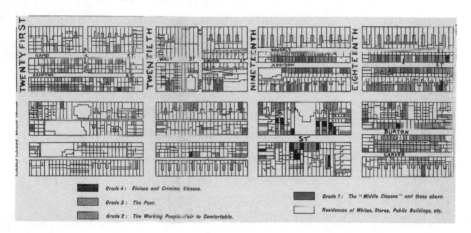

Detail from W. E. Dubois's map of Philadelphia's Seventh Ward, 1899

Those color codings themselves were a revelation even to Philadelphia progressives who lived at the boundaries of the ward, documenting the existence of a distinct class structure *within* the African-American community, one that could be clearly seen in the spatial distribution of the different categories on the map: the destitute and the criminal classes crammed together east of Seventeenth Street, the more affluent families prospering on the western boundaries of the ward, where they intermingled with white neighbors. These visualizations, and the detailed prose that accompanied them, ultimately earned *The Philadelphia Negro* the pride of place it deserves as one of the seminal early works of urban sociology, alongside the famous maps of London poverty conducted by Charles Booth in the 1880s, and Jane Addams's Hull House maps of Chicago compiled the year before Du Bois began his survey of the seventh ward. Yet Du Bois's achievement with *The Philadelphia Negro* is still largely under-appreciated, because he also made a significant advance in the science of analyzing disparities in human health. As a *sociologist*, Du Bois was working at the vanguard of his field; as a *social epidemiologist*, analyzing and explaining the disparity in health outcomes between white and black populations, he was at least a half century ahead of everyone else.

Mirroring the larger project of *The Philadelphia Negro*, Du Bois began by combating the standard-issue prejudice of the day: that there was something intrinsic to the black race that caused higher mortality rates in their community. "When attention has been called to the high death rate of this race," Du Bois explained, "there is a disposition among many to conclude that the rate is abnormal and unprecedented, and that, since the race is doomed to early extinction, there is little left to do but to moralize on inferior species. Now the fact is, as every student of statistics knows, that considering the present advancement of the masses of the Negroes, the death rate is not higher than one would expect; moreover there is not a civilized nation to-day which has not in the last two centuries presented a death rate which equaled or surpassed that of this race."[10]

Using techniques that would have impressed William Farr, Du Bois laid out the statistical evidence for the higher mortality rates of African Americans in the Seventh Ward (and greater Philadelphia) through more than a dozen charts and tables. On average, blacks were dying at a rate about 5 percent higher than that of their white neighbors. And while Du Bois never calculated life expectancy rates for the community, he did include several charts in the style of Farr's life tables that showed a shocking gap between black and white families in terms of childhood mortality. Black Philadelphians were twice as likely to die before the age of fifteen as their white neighbors.

If Du Bois had merely managed to document those racial health inequalities with such formidable statistical evidence, *The Philadelphia Negro* would have marked an important milestone in the development of "vital statistics"—building on Farr's life tables that had exposed the inequalities between city and country. But Du Bois knew he had to do more than just document the difference, given the racial prejudices of the day. More important, he had to *explain* the difference, to demonstrate that it was not just an inevitable consequence of the Seventh Ward being populated by an "inferior species." Here Du Bois brought the social into social epidemiology, using his exhaustive survey of the neighborhood's conditions to reveal the environmental causes that led to such grotesque disparities in health. And he connected those physical conditions to larger forces of discrimination—what we would now call systemic racism— at work in the city.

"Broadly speaking," he wrote, "the Negroes as a class dwell in the most unhealthful parts of the city and in the worst houses in those parts. . . . Of the 2,441 families only 334 had access to bathrooms and water-closets, or 13.7 percent. Even these 334 families have poor accommodations in most instances. Many share the use of one bathroom with one or more other families. The bathtubs usually are not supplied with hot water and very often have no water-connection at all. This condition is largely owing to the fact that the Seventh Ward belongs to the older part of Philadelphia, built when vaults in the yards were used exclusively and bathrooms could not be given space in the small houses. This was not so unhealthful before the houses were thick and when there were large back yards. To-day, however, the back yards have been filled by tenement houses and the bad sanitary results are shown in the death rate of the ward."[11]

Du Bois's in-person visits also gave him a unique vantage point for understanding the scale of the overcrowding problem in the Seventh Ward. In his survey, he documented two apartments where ten people shared a single room, and more than a hundred cases where apartments were occupied at a rate of four or more people per room. Du Bois explained how the economic realities and ingrained prejudice of the city made it inevitable that African Americans would find themselves living in such unhealthy environments. "The undeniable fact that most Philadelphia white people prefer not to live near Negroes limits the Negro very seriously in his choice of a home and especially in the choice of a cheap home. Moreover, real estate agents knowing the limited supply usually raise the rent a dollar or two for Negro tenants, if they do not refuse them altogether. . . . The mass of Negroes are in the economic world purveyors to the rich—working in private houses, in hotels, large stores, etc. In order to keep this work they must live nearby. . . . Thus it is clear that the nature of the Negro's work compels him to crowd into the center of the city much more than is the case with the mass of white working people." The fact that so many African Americans were dying early from contagious diseases like tuberculosis was not the result of some inherent propensity for disease in the "Negro race,"

Du Bois made clear; it was the indirect consequence of the way society was organized to channel African Americans into the most unsanitary spaces in the entire city. It wasn't possible to solve the health crisis of the Seventh Ward by simply demanding that African Americans adopt healthier lifestyles; the entire system had to change if we wanted those health outcomes to improve.

Like Farr and Snow's original forays into vital statistics, the innovations in data analysis that Du Bois introduced continue to play an essential role in our battle against twenty-first-century health threats. African Americans continue to lag behind white Americans in life expectancy and childhood mortality; COVID-19 has had a disproportionate impact on communities of color in the United States, in part because those communities continue to live in higher-density residences, where respiratory illnesses can easily spread. The health inequities revealed by this kind of social epidemiology has spawned a whole new field of research, exploring the way poverty and discrimination cause long-term health problems, largely through the deleterious effect of chronic stress on the body. Farr and Snow used data to reveal the way the physical infrastructure of the industrial city was cultivating disease; Du Bois took a comparable data set and connected it to the wider problem of prejudice itself.

YEARS AFTER THEIR joint investigation of the 1854 cholera outbreak, Henry Whitehead wrote that Snow had once told him: "You and I may not live to see the day and my name will be forgotten when it comes; but the time will arrive when great outbreaks of cholera will be things of the past; and it is the knowledge of the way in which the disease is propagated which will cause them to disappear." Snow was largely correct in his prophecy about the decline of cholera epidemics, though he was wrong about his name being forgotten. Today in London a replica of the pump—with a small plaque commemorating the breakthrough—stands on the sidewalk at what would have been 40 Broad Street, next to a corner pub that is now known as The John Snow. Public health workers make regular pilgrimages to the site; some sign a guestbook at the pub. But it is striking to consider the pump memorial in the light of other London tourist landmarks. So many of the grand public memorials in great cities are devoted to military events and heroes: Think of Lord Nelson towering above Trafalgar Square or the Civil War monument in Grand Army Plaza, near where I live in Brooklyn. But the pump is one of the only urban monuments I have ever seen dedicated to a public health breakthrough. And, of course, the pump memorial is to scale and almost entirely invisible unless you happen to be standing right next to it; you could easily walk along the other side of the street and not even notice the thing. Du Bois's investigations have a memorial of comparable size: a plaque in the Seventh Ward, noting that the "African-American scholar, educator, and activist" lived in the neighborhood "while collecting data for his classic study, *The Philadelphia Negro*."

There is something off kilter about the ratios here, both the number and scale of the war memorials compared to the pump on Broad Street or Du Bois's plaque in the seventh ward. To be clear, the lives lost at the Battle of Trafalgar or during the American Civil War deserve the memorials we have given them. But the pump, in a way, reminds us of a different kind of history: it is a memorial to lives *saved*, to the hundreds of thousands or millions of people who didn't die of cholera in part because a local physician in a poor neighborhood saw a pattern in the mortality data and changed our understanding of epidemic disease. (And because a statistician and a priest helped make that pattern visible.) The history of the last two centuries is filled with comparable triumphs, breakthroughs that shape our day-to-day existence in incalculable ways, particularly in large metropolitan areas where epidemic diseases were a daily reality just a few generations ago. Why not celebrate those triumphs as visibly as we celebrate the military victories?

The lopsided nature of the number of these monuments is mirrored in a more material imbalance: the difference in funding between public health institutions and the military. The United States spends about $8 billion a year on the institution that descends directly from Farr and Snow's pioneering work: the Centers for Disease Control and Prevention (CDC). The US military, by contrast, spends almost twice that amount just on space-based defense systems. The total spending for national defense is almost one trillion dollars. As I write this, the number of Americans who have died from the coronavirus in six months is more than half the total American casualties in all the wars of the twentieth century. The pandemic has made it clear that we face far greater threats from microbes than we do from our human antagonists. And the vast number of lives saved thanks to vital statistics and related public health interventions remind us that it is organizations like the WHO or the CDC that have historically done the most important work in keeping us safe.

There is a perceptual problem intrinsic to valuing that work. It did not manifest itself in the visible icons of modernity: factories, skyscrapers, rockets. Instead, it came from elsewhere, literally out of sight: in the reduction of invisible microorganisms in our drinking water, in sewer lines built below ground, in obscure publications of tabulated data. The fact that these achievements are difficult to see—and thus under-represented in our memorials and in our government spending—should not be an excuse to keep our focus on the fighter jets and the nuclear weapons. Instead, it should inspire us to correct our vision.

4

SAFE AS MILK

PASTEURIZATION AND CHLORINATION

n May 1858, not long after John Snow discovered the source of the cholera epidemic in that Soho well, a progressive journalist in Brooklyn named Frank Leslie published a five-thousand-word exposé, denouncing another brutal killer haunting the streets of a major metropolis. The piece pulled no punches in establishing the scale of the crime it was documenting: certain malevolent figures were responsible for the deaths of countless children, what Leslie called "the wholesale slaughter of the innocents." "For the midnight assassin," he thundered, "we have the rope and the gallows; for the robber the penitentiary; but for those who murder our children by the thousands we have neither reprobation nor punishment."[1]

This was the *Gangs of New York* era, so one might naturally have assumed that Leslie was denouncing the criminal underworld. His frequent references to "liquid poison" might have suggested that his was one of many temperance screeds published during that period bemoaning the social destruction of alcohol. But the mass killers that Leslie actually had in his crosshairs seem more than a little incongruous to the modern reader. Leslie was not railing against mobsters, or drug peddlers. He was denouncing milkmen.

Milk is so thoroughly associated with health and purity in the modern world that it is hard to imagine that it was, not so very long ago, one of the primary drivers of childhood mortality, potentially as deadly as the contaminated water supplies that brought cholera to so many cities around the world. In the middle of the nineteenth century, as New York experienced runaway population growth, the childhood mortality rate approached 50 percent, roughly comparable to the carnage that Farr had documented in industrial Liverpool. At the start of the century, in most large American cities, a quarter of all reported deaths involved children under the age of five, still a shocking number by modern standards. But by the 1840s, more than half of all deaths in New York were infants and young children. Something in the city was indeed "slaughtering the innocents," as Leslie put it—and seemingly at an accelerating rate. Some of those deaths were attributable to waterborne disease, particularly cholera, concentrated in terrible epidemics that laid siege to the city in 1832 and 1849. But in other years, the primary killer appears to have been contaminated milk. And while its victims were overwhelmingly children, many adults were numbered among the death toll as well. In 1850, after laying the cornerstone for the Washington Monument, the twelfth president of the United States, Zachary Taylor, died in office after drinking what many believe was a contaminated glass of milk.

Drinking animal milk—a practice as old as animal domestication itself—has always presented health risks, either through infections passed down from the animal itself or from spoilage. But a confluence of events in the first decades of the nineteenth century made cow's milk far deadlier than it had been in earlier times. Thanks to its Dutch roots, the island of Manhattan had a long tradition of dairy farmers producing milk for New Yorkers clustered at the southern tip of the island and on farms scattered across the still-rural areas in northern Manhattan and in Brooklyn. But as the city swiftly colonized those regions during the nineteenth century, traditional farmland disappeared. In an age without refrigeration, milk would spoil in summer months if it was brought in from far-flung pastures in New Jersey or upstate New York. Enterprising dairy producers recognized that they could maintain large herds of cattle in the city as long as they could figure a way to feed them without access to the wide-open pastures of Dutch-era Manhattan. They soon settled on a seemingly ingenious partnership with neighboring distilleries. The process of extracting alcohol to make whiskey from grain-generated waste products that went by several names, all of them equally unappetizing: "slop," "mash," or "swill." Instead of discarding the excess waste, the distilleries could sell it to the milk producers, who could feed it to their cows in lieu of more-expensive grain or grasses. Cows living off a diet of whiskey swill produced an unappetizing, blue-colored milk, but at least it could be delivered fresh to the exploding population of Manhattan.

The market for cow's milk had also been enhanced by the shift in labor patterns introduced by industrialization. With more women joining the workforce, breastfeeding young children beyond the first months of infancy became increasingly rare. One health expert opined that "the wear and tear of modern life, with its demands upon the mother's nervous strength and upon her time, and other factors less definitely recognized, have made it impossible for the human race to offer its progeny the sustenance intended by nature."[2] With a growing demand for cow's milk, and a symbiotic partnership with the brewers that reduced costs, whole neighborhoods of New York City were soon overrun with industrial dairy producers, with thousands of cows crowded into stalls, housed in fully urban neighborhoods in Manhattan and Brooklyn. The cows would be tied to a single stall for their entire lives, while boiling slop from the distilleries was poured into a trough in front of them. Feeding the cows exclusively swill—the dairy producers even withheld water from the animals, thinking that there was sufficient water in the distillery waste—triggered ulcerated sores and caused their tails to fall off. Many cows lost their teeth. But as gruesome as the process was, it did manage to produce copious amounts of cheap milk, which the dairy producers adulterated with chalk, flour, and eggs to make it look more like "Pure Country Milk"—the misleading branding they used to describe the product. The combination of the advertising and the cheap prices—as little as six cents per quart—soon had the working classes of Manhattan and other cities around the country hooked on swill milk. And almost immediately, children began dying at a terrifying rate.

THE STORY OF how we transformed milk from a mass killer into an emblem of health and nutrition gives us an object lesson in how we often misunderstand—or simplify beyond recognition—the factors that drive long-term improvements in human health. To begin with, most of us have a kind of historical amnesia when it comes to surprisingly recent threats like swill milk (or the contaminated drinking water that John Snow and William Farr helped purify). Most New Yorkers today have no sense that just three or four generations ago, their city-dwelling ancestors had a meaningful chance of dying in childhood because they drank a glass of milk. We remember the casualty rates of military conflicts from the period—the Civil War most prominently—because those events concentrated the killing in episodes of sudden violence. But the steady, incremental losses of children dying one by one in the slums of the industrial city do not stay fixed in our historical memory.

Even if we do manage to recall how deadly milk was in the nineteenth century, our default explanation for what removed that curse also distorts the actual history. We condense a complex network of agents into a single heroic scientist. In this case, the scientist is so prominent that his name is imprinted on the vast majority of milk cartons sold today: Louis Pasteur. Milk was once deadly; now it is safe. How did that happen? Ask people to answer that question and they will invariably tell you that pasteurization was responsible for the change. We turned milk from a liquid poison into a life-sustaining staple, thanks to chemistry.

That explanation is not so much wrong as it is woefully incomplete. One simple measure of why it is incomplete is how long it took for Pasteur's idea to actually have a meaningful impact on the safety of milk. In 1854, at the age of thirty-two, Pasteur took a job at the University of Lille in the northeast corner of France, just west of the French-Belgian border. Sparked by conversations with winemakers and distillery managers in the region, Pasteur became interested in the question of why certain foods and liquids spoiled. Knowing its propensity for spoilage, Pasteur initially focused his investigations on milk, but eventually turned to beer and wine. Examining samples of a spoiled beetroot alcohol under a microscope,

Pasteur was able to detect not only the yeast organisms responsible for fermentation, but also a rod-shaped entity—now called *Aceto-bacter aceti*—that converts ethanol into acetic acid, the ingredient that gives vinegar its sour taste. These initial observations convinced Pasteur that the mysterious changes of both fermentation and spoilage were not the result of spontaneous generation—simple chemical reactions between enzymes—but rather were the by-product of living microbes. That insight would eventually help provide the foundation of the germ theory of disease, but it also led Pasteur to experiment with different techniques for killing those microbes before they could cause any damage. By 1865, now a professor at the University of Paris, Pasteur had hit upon the technique that would ultimately bear his name: by heating it to around 130 degrees Fahrenheit, wine could be successfully prevented from spoiling without affecting its flavor in a detectable way.[3]

Today all pasteurized milk is produced using the basic technique Pasteur identified in 1865. (The temperatures have been fine-tuned over the years to be a little higher than Pasteur employed on wine.) And yet in the United States, pasteurization did not become a standard practice in the milk industry until 1915, a full fifty years after Pasteur developed the technique. The lag from discovery to implementation might well have cost millions of lives around the world. The lag happened because progress is not merely the result of scientific discovery. It also requires other forces: crusading journalism, activism, politics. Science alone cannot improve the world. You also need struggle.

THERE IS A tendency in many accounts of progress in the modern age to place the bulk of the emphasis on scientific or technical breakthroughs, and largely ignore the agitators and muckrakers and political coalitions that drove the improvements in public health over the past two centuries. The battle against cholera gives us a good case study in this neglected aspect of positive change: John Snow was, technically speaking, a physician, one who became famous for his work as an epidemiologist. But surely part of the story of the Broad Street pump is that Snow had to fight for his ideas in the political arena as well: petitioning the local parish board and citywide board of health to change its approach to cholera. The triumphs over waterborne disease in the nineteenth century were the result of social movements as much as they were triumphs of Enlightenment science. That's one reason a figure like Henry Whitehead was able to play such an important role, despite the fact that he lacked any scientific or medical training. He had social capital in the community that ultimately helped change the minds of the authorities.

In an artful critique of Thomas McKeown's work on mortality decline, the Cambridge historian Simon Szreter has argued for the importance of what Szreter called social intervention between 1850 and the outbreak of World War I. Yes, Szreter argued, it was true that diseases like cholera had abated during that period despite the fact that no medicines had been discovered that would properly treat them. Miracle drugs didn't cure cholera. Sewers did. And sewers were government-funded projects that only came into being because campaigners like Snow and Bazalgette—along with their advocates in the popular press—argued for their existence:

> The decline in mortality, which began to be noticeable in
> the national aggregate statistics in the 1870s, was due more
> to the eventual successes of the politically and ideologically

negotiated movement for public health than to any other positively identifiable factor. This was only achieved as a result of innumerable unsung local skirmishes between frequently underpaid health officials, often lacking security of tenure, and their local allies—other sanitary officials, the district registrars of births and deaths, perhaps the town's press and occasionally some members of the local councils themselves—as against the parsimonious representatives of the majority of ratepayers. It is precisely the importance and necessity of this slow dogged campaign of a million Minutes, fought out in town-halls and the local forums of debate all over the country over the last quarter of the nineteenth century which has been missing in our previous accounts of the mortality decline.[4]

The battle for safe milk gives us an even more compelling illustration of Szerter's social interventions at work. The first volley in that battle arrived in the early 1840s, as the swill milk establishments were proliferating across New York City, in the form of a book published by a dry goods merchant named Robert Milham Hartley. As one of the founders of the New York temperance movement, Hartley was predisposed to detect the pernicious influences of breweries and distilleries in the city, while his missionary work as a member of the Presbyterian Church had given him firsthand experience with the appalling living conditions in the slums of Five Points. Anecdotal evidence from that missionary work suggested that there had been a disturbing spike in childhood deaths, which he then confirmed through a careful study of the city's mortality reports. Hartley began a private investigation into the dairy producers, ultimately publishing a 350-page book with the baroque title, *An Historical, Scientific, and Practical Essay on Milk: As an Article of Human Sustenance; with a Consideration of the Effects Consequent Upon the Present Unnatural Methods of Producing it for the Supply of Large Cities.* The book combined data analysis in the mode of William Farr's annual reports—including a comparative study of childhood mortality rates in American and European cities—along with vivid journalistic accounts of the scandalous conditions at the slop dairies:

> One of the most notorious of these overgrown metropoli
> tan milk-establishments, or rather the largest collection of
> slop-dairies, for there are many proprietors, is that con-
> nected with Johnson's grain-distilleries, which are situated
> in the western suburbs of the city, near the termination,
> and between Fifteenth and Sixteenth streets, in New-York.
> The area occupied by the concern, includes the greater part
> of two squares, extending from below the Ninth Avenue to
> the Hudson River, probably a distance of one thousand
> feet. During the winter season, about two thousand cows
> are said to be kept on the premises, but in summer the
> number is considerably reduced. The food of the cows, of
> course, is slop, which being drawn off into large tanks, el-
> evated some ten or fifteen feet, is thence conducted in close
> square wooden gutters, and distributed to the different
> cowpens, where it is received into triangular troughs, rudely
> constructed by the junction of two boards.[5]

In the ultimate tally, Hartley concluded that "about ten thousand cows in the city of New York and neighborhood are most inhumanely condemned to subsist on the residuum or slush of this grain, after it has undergone a chemical change, and reeking hot from distilleries."

The *Essay on Milk* was a damning indictment of the entire dairy industry, but for some reason it failed to sway popular opinion or inspire government intervention. Part of that is because the government lacked suitable regulatory bodies to deal with the swill milk crisis, most of which would not be invented until the twentieth century (see chapter 5). But part of Hartley's failing seems to have been caused by his prominence in the temperance movement. The city was in the middle of a decades-long bender, and it didn't have the patience for a sermon denouncing the West Village distilleries from a teetotaling missionary.

In the end, it was Frank Leslie's epic renunciation of swill milk's liquid poison, published a decade and a half after Hartley's essay, that finally provoked meaningful reform. The tone of both works are similar in many places: righteous indignation interspersed with almost prurient descriptions of dairy industry atrocities. "Though their traffic is literally in human life," Leslie wrote of the swill milk impresarios, "the Government seems powerless or unwilling to interfere. . . . Shall these manufactories of hell-broths be permitted longer to exist among us? Shall we tamely submit that a class of men shall grow rich upon our bereavements—upon the vacant places their poison creates in every family?"[6] But Leslie fought his battle with more than just words. Originally trained as an illustrator, Leslie peppered his investigative report with shocking images that conveyed the squalor of the swill milk plants. (Some of them were drawn by the legendary illustrator Thomas Nast.) One illustration showed a diseased cow no longer able to stand, suspended in the air by straps, its head hung low as if barely conscious. Despite the animal's appalling physical condition, a dairy worker sits on a stool, dutifully extracting milk from the cow's ulcerous udders.

Anti-swill milk cartoon from printers Currier and Ives, 1872

Leslie's first big break had come designing promotions for P. T. Barnum, and he approached his work as a muckraking journalist with a Barnumesque flair for publicity. He took out ads in rival newspapers promoting his special report with provocative headlines that sounded like a tease for the eleven o'clock news broadcasts of today's world: ARE YOU AWARE OF WHAT KIND OF MILK YOU ARE DRINKING? Before long, Leslie's investigative report had become its own story. The *New York Times* wrote of Leslie's efforts:

> Unpromisingly matters stood, when Frank Leslie found
> left at his door as milk a disgusting dose of milk and pus,
> which fairly threw his illustrated newspaper into an emetic

convulsion. Bound to know the worst of the horrible story, he analyzed the specimen, and then dispatched his corps of reporters and artists to the headquarters of the poison. . . . He has reproduced pictures that are true to the life, and so shocking that the very word milk, or the sight of dainties into which it enters as an important component, turns the stomach. The whole town suffers nausea.[7]

The Times's account of the investigation's origins is more myth than reality: Leslie's original inspiration for his exposé was not a bottle of spoiled milk on his doorstep but rather a report that had been commissioned by the Common Council of Brooklyn a year earlier, which documented the rampant animal abuse in the swill milk dairies. But whatever its origins, Leslie's gift for promotion—and the genuinely appalling facts of the case—led to meaningful reform. Pressured by Leslie's journalism, the Common Council launched an investigation of the swill milk dairies, but then issued a quick ruling that proposed only modest changes, no doubt because the council members were secretly being paid off by the milk industry. Leslie countered three days later with a satiric Thomas Nash illustration that depicted one Tammany Hall politician receiving a bribe from a milk magnate, while his colleagues literally whitewashed the dying cattle to make them appear healthier. Popular outrage made it impossible for the council to avoid real action, and by 1862, legislation had been passed that put an end to the swill milk era. Most of the urban dairies shut down; those that remained gave up their sordid partnerships with the distilleries. The milk of New York shed that strange blue color.

Still, many health risks from milk consumption remained. With more milk traveling from upstate farms, spoilage continued to be a serious risk, particularly in the summer months. And a significant portion of milk-producing cows—even the ones on proper dairy farms in the country—suffered from bovine tuberculosis. Unprocessed milk from these cows could transmit the tuberculosis bacterium to human beings. Other potentially fatal illnesses were also linked to milk, including diphtheria, typhoid, and scarlet fever. Frank Leslie's campaign had shown that public opinion could be mobilized to reform the milk industry. But swill milk was only part of the problem.

BY THE 1880S, the emergence of the germ theory of disease—the seeds of which had been planted in Pasteur's early research into milk and wine spoilage—had made it clear that many of the century's deadliest killers were caused by microbial life-forms, newly visible thanks to advances in lens making that enabled more powerful microscopes. In 1882, Pasteur's archrival Robert Koch identified the tuberculosis bacterium, disproving a long-standing belief that the disease was inherited; two years later, he identified the cholera bacterium that had eluded John Snow's microscopic investigations decades before. The science was settled: people were dying from drinking milk because it contained invisible creatures that caused disease. But that consensus left a further problem unsolved: How do you keep those creatures out of the milk supply?

One critical part of the solution would come from technological innovation. In the first half of the nineteenth century, the tenacious Boston-based entrepreneur Frederic Tudor had built an immense business selling ice all around the world, ice that was used directly by consumers in their drinks and ice cream, but also by the food industry to refrigerate transport, most famously allowing meat from the Great Plains to feed the growing cities of the Northeast. Tudor employed low-tech methods of creating ice—he had teams carving it out of frozen New England lakes—but his vast fortune sent a signal out to inventors everywhere that there was money to be made in making things cold. By the end of the Civil War, a number of functional designs for mechanical refrigeration had been developed, and by the end of the century, milk bottles could be stored and shipped in temperature-controlled environments, greatly reducing the risks of spoilage. Refrigeration turns out to be one of those underlying technologies that doesn't seem to be directly related to medicine, but that advances public health and longevity on multiple fronts. Its ability to prolong the shelf life of perishable foods had a tremendous impact on food supply in the twentieth century, and it helped turn milk from a liquid poison to a reliable source of nutrition. But refrigeration also had a critical impact on vaccines, many of which lose their potency if they are not maintained in a narrow band of temperatures just above freezing. The invention of "cold-chain" supply networks enabled mass vaccinations in many hot climates around the world where diseases like smallpox remained endemic for much of the twentieth century.

But refrigeration was only a partial solution. A carton of milk contaminated with bovine tuberculosis could still be deadly, even if it had spent its entire existence in a refrigerator. For some milk reformers, the obvious solution was to follow the playbook that had worked in getting rid of the swill milk dairies: attack the problem at its source. Tests had been developed that could determine whether a cow was suffering from diseases like bovine tuberculosis; new microscopes enabled scientists to analyze milk to determine the quantity of bacteria it contained. Armed with these new tools, milk inspectors could potentially visit dairy producers and certify that the cows were free of disease and that the conditions were sanitary. Milk produced by dairies that had passed these inspections would be "certified," giving consumers confidence that the milk they were purchasing would be safe to drink.

But the certification approach had its own set of problems. Cows that had contracted bovine tuberculosis would have to be slaughtered, and rough estimates suggested that as many as half of the dairy cows in the country were carriers of the disease. Understandably, rural dairy producers resented urban inspectors showing up on the farms with some mysterious tuberculosis test and announcing that cows that seemed perfectly healthy would have to be destroyed. Politicians representing farming districts fought back against these encroachments. And until the Food and Drug Administration was formed in 1906, there was no federal body capable of enforcing such regulations.

Thanks to Louis Pasteur, however, tuberculin tests and dairy inspections were not the only tools available to the milk safety advocates. Refrigeration could keep milk from spoiling, but a quick burst of heat could kill the dangerous microbes in milk, even the ones that caused tuberculosis in humans. Once again, though, the science was not sufficient on its own to create meaningful change. Pasteurized milk was widely considered to be less flavorful than regular milk; the process was also believed to remove the nutritious elements of milk—a belief that has reemerged in the twenty-first century among "natural milk" adherents. Dairy producers resisted pasteurization not just because it added an additional cost to the production process, but also because they were convinced—with good reason—that consumers wouldn't purchase pasteurized milk.

As is so often the case in the history of modern human life expectancy, the turning point that brought the life-saving innovation to a mass audience would not involve a scientist or a doctor in a leading role. Pasteurization as an idea had first developed in the mind of a chemist. But in the United States, it would finally make a difference thanks to a much less likely character: a department store impresario.

BORN IN BAVARIA in 1848, Nathan Straus moved at the age of eight with his family to the American South, where his father had established a profitable general store. The move turned out to have been disastrously timed. Pushed to the edges of abject destitution by the Civil War, the family relocated to New York just as Nathan was reaching adulthood. In Manhattan, the Straus clan found their footing. Nathan began his career by working for his father's crockery and glassware firm; he and his brothers sold the pans and plates they manufactured to the new department stores that had exploded onto the commerce and fashion world in the 1870s. In early 1873, they began renting a space in the basement of Macy's flagship Fourteenth Street store to display their china, glass, and pottery. It soon became one of the most popular destinations in the store. A little more than a decade later, the Straus brothers had acquired Macy's outright, along with a prominent Brooklyn dry goods outlet, Abraham & Straus.

Perhaps because his own family had had a near-death experience with sudden poverty, Nathan Straus spent a significant measure of his time and resources attempting to improve the conditions of New York City's homeless and working poor. He opened shelters that housed more than fifty thousand people and distributed coal during the brutal winter and economic downturn of 1892–93. At Abraham & Straus, he built a cafeteria on the grounds that offered a free meal plan for his employees, one of the first such programs ever created. Straus had long been concerned about the childhood mortality rates in the city—he had lost two children to disease. Conversations with another German émigré, the political radical and physician Abraham Jacobi, introduced him to the pasteurization technique, which was finally being applied to milk almost a quarter of a century after Pasteur had first developed it. Something about the process resonated with Straus—given the complexities of urban poverty, pasteurization offered a comparatively simple intervention that could make a meaningful difference in keeping children alive.

Straus recognized that changing popular attitudes toward pasteurized milk was key. In 1892, he created a milk laboratory where sterilized milk could be produced at scale. The next year, he began opening what he called milk depots in low-income neighborhoods around the city that sold milk to poor New Yorkers below cost. The first depot was located on a pier on the outer edges of the Lower East Side; records suggest that Straus dispensed 34,400 bottles of milk that first year. By the summer of 1894, Straus had opened four depots around the city.[8] The *New York Times* ran a story on the new depots, headlined PURE MILK FOR THE POOR. In it Straus was quoted saying, "The success of the milk depot last Summer has led me to extend the facilities. The only trouble is that the poor do not yet fully understand the value of the sterilized milk as a remedy for the sickness of children and a preventative. I have reduced the price of the sterilized milk to 5 cents a quart, which is below the cost price. It is possible that I may reduce the price still further."[9]

Appointed health commissioner of the city in 1897, Straus learned of devastating mortality rates at an orphanage situated on Randall's Island in the East River. In the preceding three years,

1,509 of the 3,900 children housed in the orphanage had perished—
a mortality rate even higher than the dismal rates in low-income
communities around the city. Straus suspected that the dairy herd
that had been established on the island to supply the orphans with
fresh milk was in fact the culprit. He realized that the geographic
isolation of the orphanage presented a natural experiment to prove
the efficacy of pasteurized milk, not unlike the natural experiment
that John Snow had conducted during the cholera outbreak of 1854.
Straus funded a pasteurization plant on Randall's Island that supplied sterilized milk to the orphans. Nothing else in their diet or
living conditions was altered. Almost immediately, the mortality
rate dropped 14 percent.[10]

Emboldened by the results of these early interventions, Straus
launched an extended campaign to outlaw unpasteurized milk,
which was ferociously opposed by the milk industry and its representatives in statehouses around the country. Pasteurization became a political fight. Quoting an English doctor at a rally in 1907, Straus told
an assembled mass of protesters: "The reckless use of raw, unpasteurized milk is little short of a national crime."[11] Straus's advocacy
attracted the attention of President Theodore Roosevelt, who or-
dered an investigation into the health benefits of pasteurization.
Twenty government experts came to the resounding conclusion that
"Pasteurization prevents much sickness and saves many lives." In
1909, Chicago became the first major American city to require pasteurization. The city's commissioner of health specifically cited the
demonstrations of the "philanthropist Nathan Straus" in making
the case for sterilized milk. New York followed suit in 1914. By the
early 1920s, three decades after Nathan Straus opened his first milk
depot on the Lower East Side, unpasteurized milk had been outlawed in almost every major American city.[12]

. . .

THE IMPACT OF PASTEURIZATION on life expectancy is difficult to measure exactly because the mortality data is confounded by another key breakthrough from the same period—a breakthrough that also employed chemistry to reduce the threat of an everyday liquid. Starting in the first decades of the twentieth century, human beings in cities all around the world began consuming microscopic amounts of chlorine in their drinking water. In sufficient doses, chlorine is a poison. But in very small doses, it is harmless to humans but lethal to the bacteria that cause diseases like cholera. Thanks to the same advances in microscopy and lens making that had enabled bacterial counts in milk, scientists could now perceive and measure the amount of microbial life in a given supply of drinking water, which made it possible by the end of the nineteenth century to test the efficacy of different chemicals, chlorine above all else, in killing off those dangerous agents. After conducting a number of these experiments, a pioneering doctor named John Leal secretly added the chemical to the public reservoirs in Jersey City— and audacious act that got Leal in so much trouble that he was almost sent to prison. To a nonscientist, it seemed patently insane to introduce a poisonous chemical into the primary supply of drinking water for a city of fifty thousand people. But Leal's daring move turned out, in the long run, to be a lifesaver on an astonishing scale.[13]

From 1900 to 1930, infant mortality rates in the United States dropped by 62 percent, one of the most dramatic declines in the history of that most critical of measures. For every hundred human beings born in New York City for most of the nineteenth century, only sixty would make it to adulthood. Today ninety-nine of them do. Gradients continue to haunt the city: travel a few stops on the number 2 train in Brooklyn and you can easily find yourself in a neighborhood with twice the infant mortality rate as the one you started in. But even those low-income communities are staggeringly good at keeping babies alive, compared to every known human society before 1900. The change is so pronounced that it requires an extra decimal point. Nancy Howell estimated that the !Kung people had an infant mortality rate of around 20 percent; before Frank Leslie and Nathan Straus began their publicity campaigns, newborns in *Gangs of New York* Manhattan experienced similar mortality levels. Today, in the worst-performing neighborhoods of New York, the infant mortality rate is 0.6 percent. The citywide average is 0.4 percent.[14]

How much of that is attributable to pasteurization and chlorination, those two great triumphs of chemistry? In the early 2000s, the Harvard professors David Cutler and Grant Miller hit upon an ingenious approach for analyzing the effect of chlorine on mortality rates. Because these filtration techniques were introduced in a staggered fashion, with some cities adopting them before others, comparing before- and after-chlorination mortality rates between those cities gave the researchers a natural experiment of sorts. Through their comparative analysis of the different cities, Cutler and Miller determined that filtration techniques like chlorination had been responsible for more than two thirds of that dramatic improvement.[15] Their study became something of a classic among public health scholars, though in recent years, attempts to replicate their data suggest that the impact on overall mortality was not as dramatic, in part because pasteurization played such an important role as well.

However you analyze the data, it's clear that millions of newborn children made it to adulthood thanks to government-regulated chemistry, and the insurgents who fought for it. When you ask people to list the great innovations of the early twentieth century, they invariably list planes, automobiles, radio, television—not pasteurized milk or chlorinated drinking water. But think of all the unimaginable suffering that was avoided by those two interventions: all the parents who didn't bury their children, all the infants who got to grow up and have their own children in turn.

What were the ingredients behind such dramatic progress? Undeniably, there were the usual suspects: brilliant scientists like Koch and Pasteur, supported by the technical innovations of microscopy. But the agitators were essential as well. The swill milk scandal and the fight for pasteurized milk were media events as much as they were triumphs of Enlightenment science. To make our milk safe to drink, we needed a chemist using the scientific method to invent a technique that killed off the contaminants. But we also needed people willing to make some noise.

In 1908, right in the heat of the battle over the first proposed ordinance banning unsterilized milk, Nathan Straus was invited to give an address at the University of Heidelberg, returning to the country his family had fled more than fifty years before. In the talk, he raised the question of why he had become so invested in the cause of pasteurization, briefly alluding to the trauma losing a child (or in Straus's case, two children) inflicts on a parent. "Into the personal and private reasons that first induced me to engage in this work I need not enter here," he told the crowd. "It is enough to say that it was my own sad experience which made me so determined to save the lives of other people's babies." But then he turned to methods he had employed in his struggle. "I have always only considered how best and quickest to enlighten the world in a practical manner. To attain this I sought the help of the press, and it is due to its ever ready cooperation that my work and its results have been made known and broadcast. Only through publicity can the advantages of the pasteurization of milk be everywhere realized."[16]

THE WESTERN DECLINES in infant mortality would not reach the developing world for another half century or so. But when they eventually rolled in, they quickly made up for lost time. India's infant mortality rate dropped from 14 percent to 3 percent between 1970 and the present day. Techniques like pasteurization and chlorination played a role in that decline. But their effect may have been eclipsed by another breakthrough that greatly reduced the risk of waterborne disease, particularly cholera. As it happens, this breakthrough also relied on a hybrid strategy of chemically treated liquids combined with creative public relations.

Cholera kills by creating acute dehydration and electrolyte imbalance—caused by severe diarrhea—in those unlucky enough to have ingested the bacterium. In some extreme cases, cholera victims have been known to lose as much as 30 percent of their body weight through expelled fluids in a matter of hours. As early as the 1830s, doctors had observed that treating patients with intravenous fluids could keep them alive long enough for the disease to run its course; by the 1920s, treating cholera victims with IV fluids became standard practice in hospitals. But by that point, cholera had become a disease that was largely relegated to the developing world, where hospitals or clinics and trained medical professionals were scarce. Setting up an IV for patients and administering fluids was not a viable intervention during a cholera outbreak affecting hundreds of thousands of people in Bangladesh or Lagos. Crowded into growing cities, lacking both modern sanitation systems and access to IV equipment, millions of people—most of them small children—died of cholera over the first six decades of the twentieth century.

The sheer magnitude of that loss was a global tragedy, but it was made even more tragic because a relatively simple treatment for severe dehydration existed, one that could be performed by nonmedical professionals outside the context of a hospital. Now known as oral rehydration therapy, or ORT, the treatment is almost maddeningly simple: give people lots of water to drink, supplemented with sugar and salts. (In the United States, the treatment is often associated with the brand Pedialyte.) As early as 1953, an Indian doctor named Hemendra Nath Chatterjee had improvised a version of this therapy while treating patients during an outbreak in Calcutta. No complicated and expensive IV setup was required. All you needed was a method for ensuring the water was sterilized; boiling it before drinking would do. The results of Chatterjee's therapy were so promising—all 186 patients treated with this method survived their illness—that he published his results in *The Lancet*.[17] Other similar approaches were developed over the following decade in the Philippines and in Iraq, each developed by doctors, like Chatterjee, scrambling to deal with an explosive outbreak without recourse to the high-tech equipment of a modern hospital. And yet all these versions of ORT were ignored by the medical establishment, just as the miasma theorists had ignored John Snow's heretical waterborne theory a century before.

In 1971, the Bangladesh Liberation War sent a flood of refugees into India that swelled in cities such as Bangaon and Kolkata, located just across what would become the border between India and Bangladesh, once that nation was formally recognized after the war of independence ended. Before long, a vicious outbreak of cholera arose in the crowded refugee camps outside of Bangaon. A Johns Hopkins–educated physician and cholera researcher named Dilip Mahalanabis suspended his research program in a Kolkata hospital lab and immediately went to the front lines of the outbreak. Mahalanabis later recalled the sheer scope of the crisis: "The government was unprepared for the large numbers. There were many deaths from cholera, many horror stories. When I arrived, I was really taken aback."[18] The most shocking scene was one he encountered in a Bangaon hospital: two rooms filled wall to wall with cholera victims in the throes of the disease, lying against each other on the floor, which was itself coated in a layer of watery feces and vomit.

Mahalanabis quickly realized that the existing IV protocols were not going to work. Only two members of his team were even trained to deliver IV fluids. "In order to treat these people with IV saline," he later explained, "you literally had to kneel down in their feces and their vomit. Within forty-eight hours of arriving there, I realized we were losing the battle."[19]

And so Mahalanabis decided to shake things up. Going against standard practice, he and his team turned to an improvised version of oral rehydration therapy. He delivered it directly to the patients he had contact with, like those sprawled bodies on the floor of the Bangaon hospital. Under Mahalanabis's supervision, more than three thousand patients in the refugee camps received ORT therapy. The strategy proved to be an astonishing success: mortality rates dropped by an order of magnitude, from 30 percent to 3 percent, all by using a vastly simpler method of treatment.

Inspired by the success, Mahalanabis and his colleagues adopted a "teach a man to fish" approach, with field-workers demonstrating how easy it was for nonspecialists to administer the therapy themselves. "We prepared pamphlets describing how to mix salt and glucose and distributed them along the border," Mahalanabis later recalled. "The information was also broadcast on a clandestine Bangladeshi radio station."[20] Boil water, add these ingredients, and force your child or your cousin or your neighbor to drink it. Those were the only skills required. Why not let amateurs into the act?

In 1980, almost a decade after the end of the War of Liberation, a Bangladeshi nonprofit known as BRAC devised an ingenious plan to evangelize the ORT technique among small villages throughout the young nation. A team of fourteen women, accompanied by a cook and a single male supervisor, traveled from village to village, demonstrating how to administer oral saline using only water, sugar, and salt. The pilot program generated encouraging results, and so the Bangladeshi government replicated it on a national scale, employing thousands of field-workers. "Coaxing villagers to make the solution with their own hands and explain the messages in their own words, while a trainer observed and guided them, achieved far more than any public-service ad or instructional video could have done," the physician and author Atul Gawande wrote of the project. "Over time, the changes could be sustained with television and radio, and the growth of demand led to the development of a robust market for manufactured oral rehydration salt packets."[21] Deaths from cholera and other intestinal diseases have plummeted, and one survey suggested that 90 percent of children experiencing severe diarrhea in Bangladesh are now treated with ORT.

The Bangladeshi triumph was replicated around the world. ORT is now a key element of UNICEF's program to ensure childhood survival in the global south, and it is included on the World Health Organization's List of Essential Medicines. *The Lancet* called it "potentially the most important medical advance of the 20th century." As many as fifty million people are said to have died of cholera in the nineteenth century. In the first decades of the twenty-first century, less than fifty thousand people died of cholera, on a planet with more than ten times the population. That momentous leap forward was partly due to the nineteenth-century data detectives, the sewer engineers, and John Leal's chlorinated water. But ORT played a crucial role as well, particularly at the end.

Why did ORT take so long to enter the mainstream of medical thought? In part, the treatment took so long to circulate because of a kind of institutional bias: major discoveries were not supposed to come from doctors working in the field in countries like India or Iraq. "In the 1950s," the medical historian Joshua Nalibow Ruxin writes, in a definitive study of ORT's history, "the physiological paradigm under which Western physicians operated was that intravenous therapy was superior to all others. Thus, a researcher who read the study by Chatterjee might have thought that the concept was interesting but that Western medicine had surpassed any simplistic (and therefore inferior) solutions to cholera. Intravenous therapy appeared more scientific, there was an apparatus, and the physician could have precise control over the intake of a patient. Oral therapy appeared primitive and less controlled."[22]

ORT also arrived so late because the phenomenon it relied upon was not truly understood scientifically until the mid-1960s, when a number of researchers working in labs around the world finally determined the specific mechanism by which the cholera bacterium induced such massive fluid loss. They also discovered that glucose could promote fluid absorption in the small intestine. It was easier to promote a treatment whose underlying mechanics had the imprimatur of scientific studies, even if the evidence for the benefit of the treatment had been readily available for two decades.

There is an interesting symmetry between the story of pasteurization and the painfully slow adoption of ORT. The tipping point for both breakthroughs emerged out of medical crises—all those children dying in the orphanage on Randall's Island and in the Bangladeshi refugee camps. Both relied on inventive strategies to get the word out: Straus's milk depots, Mahalanabis's pamphlets. And in both cases, it took far too long to implement the original advance. Both milk pasteurization and ORT could have become mainstream practices a generation or more earlier than they did. In both cases, we can celebrate the achievement and marvel at all the lives—particularly the lives of young children—that were saved by such extraordinary collaborations. But we should also ask the hard questions: Why did they take so long? And what equivalent blind spot are we living with today?

5

BEYOND THE
PLACEBO EFFECT

DRUG REGULATION AND TESTING

Why do some innovations take longer than they should to arrive? For understandable reasons, the story of social and intellectual progress is usually presented as a ladder of clearly defined steps, each transformative idea supplying the foothold for the next. The rare occasions where an idea appears to skip a step or two—like Charles Babbage inventing the programmable computer in the 1830s—are the exceptions that prove the rule of linear progress. We have a term for those kinds of ideas—they're "ahead of their time." But we don't spend enough time examining the laggards, the weird gaps in the fossil record where a good idea that was clearly imaginable at a certain point in history somehow stayed out of reach. The ideas that were, somehow, *behind* their time. No one need wonder why gene sequencing wasn't invented in the late nineteenth century—both the tools and the concepts for even imagining such an advance simply didn't exist then. But we *should* wonder why something like oral rehydration therapy didn't take root fifty years before it became a mainstream practice. The idea was well within the boundaries of existing scientific understanding in that period. But for some reason, we weren't able to see it.

Technological history features a number of these puzzling laggards. Typewriters, for instance, weren't invented until the 1860s, five hundred years after Gutenberg invented the printing press. Bicycles didn't become commercially viable until the same period, just a few decades before the invention of the automobile. Many advanced civilizations failed to invent the wheel despite its simplicity and clear utility. All these ideas could have become part of technological reality much earlier than they did; there was no obvious conceptual or mechanical missing foothold on the ladder to keep us from taking the step. And yet for some reason we took centuries to take it.

There are equivalent late arrivals on a macro level, broader developments that should have been achievable but for some reason took a strangely long time to emerge. One of those laggards—as the work of Thomas McKeown revealed—is the discipline of medicine itself.

If you happened to be a pharmacist in 1900 looking to stock your shelves with medicinal cures for various ailments—gout, perhaps, or indigestion—you would likely consult the extensive catalog of Parke, Davis & Company, now Parke-Davis, one of the most successful and well-regarded drug companies in the United States. In the pages of that catalog, you would have seen products like Damiana et Phosphorus cum Nux, which combined a psychedelic shrub and strychnine to create a product designed to "revive sexual existence." Another elixir by the name of Duffield's Concentrated Medicinal Fluid Extracts contained belladonna, arsenic, and mercury. Cocaine was sold in an injectable form, as well as in powders and cigarettes. The catalog proudly announced that the drug would "[take] the place of food, make the coward brave, the silent eloquent and . . . render the sufferer insensitive to pain." As the medical historian William Rosen writes, "Virtually every page in the catalog of Parke, Davis medications included a compound as hazardous as dynamite, though far less useful."[1]

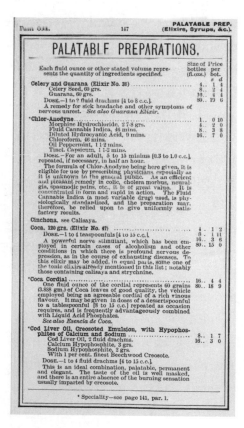

Recommended medicines in the 1907 Parke, Davis & Company catalog

This was the sad state of medicine at the beginning of the new century. Electricity had been tamed and was being used to light up the streets of Manhattan; the human race was on the verge of cracking the mystery of flight; radio signals were being transmitted through the ether. But where medicine was concerned, one of the most valuable drug companies in the world was still selling bogus cures based on mercury and arsenic.

It is likely that *personal* health interventions—as opposed to the public ones, such as sewers and water filtration systems—did not have a meaningful effect on human life expectancy until 1950. Vaccines had saved many lives over the preceding century, it was true, but the rest of the field of medicine had barely advanced from the mercury poisoning that was used to treat the mad king George. Added all together—the lives extended versus the lives shortened— the medical profession barely broke even. The historian John Barry notes that "the 1889 edition of the Merck Manual of Medical Information recommended one hundred treatments for bronchitis, each one with its fervent believers, yet the current editor of the manual recognizes that 'none of them worked.' The manual also recommended, among other things, champagne, strychnine, and nitroglycerin for seasickness." Oliver Wendell Holmes famously quipped, "I firmly believe that if the whole materia medica [medical drugs], as now used, could be sunk to the bottom of the sea, it would be all the better for mankind—and all the worse for the fishes."[2] Holmes wrote the lines in 1860, but they were almost as applicable to the state of medicine in the early twentieth century.

Today, of course, we think of medicine as one of the pillars of modern progress, alongside smartphones and electric cars. Antibiotics treat many of the illnesses that killed our great-grandparents' generation; miraculous new immunotherapies are curing cancers;

antiretroviral drugs can now effectively stop AIDS in its tracks. But those miracle drugs are actually a remarkably recent invention. Just eighty years ago, before the outbreak of World War II, the overwhelming majority of medicine on the market was useless, if it wasn't actively harmful. There is something strangely asynchronous about medicine's sorry state in the first half of the twentieth century. What was holding back the science of medicine when so many other fields were climbing the ladder of progress?

Several important factors explain medicine's late arrival. But one of the most critical has to be that there was no legal prohibition on selling junk medicine. In fact, the entire pharmaceutical industry was almost entirely unregulated for the first decades of the twentieth century. Technically speaking, there was an organization known as the Bureau of Chemistry, created in 1901 to oversee the industry. But this initial rendition of what ultimately became the US Food and Drug Administration was toothless in terms of its ability to ensure that customers were receiving effective medical treatments. Its only responsibility was to ensure that the chemical ingredients listed on the bottle were actually present in the medicine itself. If you wanted to put mercury or cocaine in your miracle drug, the FDA had no problem with that as long as you mentioned it on the label.

It took a national tragedy to change that preposterous state of affairs. In the early 1930s, the German drug company Bayer AG developed a new class of drug called sulfanilamides, or "sulfa" drugs, a less-effective forerunner of modern antibiotics. Within a few years, the market was flooded with copycat medicines. Unfortunately, sulfanilamide was not soluble in either alcohol or water, so the existing sulfa drugs came in the form of pills that were particularly challenging for children to swallow. Sensing a market opportunity, a twenty-seven-year-old Tennessean named Samuel Evans Massengill dropped out of medical school to start his own drug company with the aim of producing a sulfa variant that would be easier to consume. In 1937, the chief chemist Harold Watkins at the newly formed S. E. Massengill Company hit upon the idea of dissolving the drug in diethylene glycol, with raspberry flavoring added to make the concoction even more palatable to children. The company rushed the concoction to market under the brand Elixir Sulfanilamide, shipping 240 gallons of the medicine to pharmacies around the United States, promising a child-friendly cure for strep throat.[3]

While sulfa did in fact have meaningful antibacterial effects, and the raspberry flavoring added the proverbial spoonful of sugar, diethylene glycol is toxic to humans. Within weeks, six deaths were reported in Tulsa, Oklahoma, linked to the "elixir," each one from kidney failure. The deaths triggered a frantic nationwide search, with agents from the Food and Drug agency poring over pharmacy records, alerting doctors, and warning anyone who had purchased the drug to immediately destroy it. But the FDA didn't have enough pharmacological expertise on staff to determine what made the drug so lethal. And so they outsourced that detective work to a South Africa–born chemist at the University of Chicago named Eugene Geiling. Within weeks, Geiling had his entire team of graduate students testing all the ingredients of the elixir on a small menagerie of animals in the lab: dogs, mice, and rabbits. Geiling quickly identified diethylene glycol—a close chemical relative of antifreeze—as the culprit.

It was an inspiring combination of fieldwork and lab analysis. But for many families around the United States, it came too late. By the time the FDA recovered the last bottle, seventy-one adults and thirty-four children had died from consuming the elixir. Many more had been hospitalized with severe kidney problems, narrowly avoiding death.

Amazingly, at that moment in American history, the government still lacked a cabinet-level position that had direct oversight of the nation's health. (The US Department of Health, Education, and Welfare was not created until 1953.) And so management of this deadly drug crisis fell to Henry Wallace, then secretary of agriculture. Hauled before Congress to explain how such a lethal elixir had made it into consumers' hands, Wallace explained how the FDA had followed its oversight. "Before the elixir was put on the market, It was tested for flavor but not for its effect on human life," Secretary Wallace later reported to Congress. "The existing Food and Drugs Act does not require that new drugs be tested before they are placed on sale."[4] The examiners at the FDA had confirmed that Elixir Sulfanilamide tasted like raspberries as advertised. They just didn't bother to investigate whether it caused kidney failure.

TRAGEDIES LIKE THE ELIXIR Sulfanilamide case inevitably produce a search for villains and scapegoats, the evildoers responsible for the deaths of innocent children. No doubt part of the blame for the tragedy fell on Harold Watkins and the S. E. Massengill Company. Massengill was ultimately fined $24,600 for selling the poison to unwitting consumers, despite publicly denying Watkins's culpability. "We have been supplying a legitimate professional demand and not once could have foreseen the unlooked-for results," he declared. "I do not feel that there was any responsibility on our part."[5]

Harold Watkins, the chemist, could not brush off his responsibility for the tragedy so easily. He committed suicide before the FDA investigation was complete.

Yet it is too simple to reduce the Elixir Sulfanilamide case down to the actions of a few malevolent individuals. Those 105 deaths were also the result of both market and regulatory failures. The problem did not just lie in one rogue chemist and a reckless entrepreneur. The problem also involved the entire system of how drugs were created and sold. The pharmaceutical companies had no legal incentive to concoct elixirs that actually *worked*, given the limited oversight of the FDA. As long as their lists of ingredients were correct, they had free rein to sell whatever miracle potion they wanted. Even when one of those ingredients happened to be a known poison that killed 104 people, the penalty was only a financial slap on the wrist.

One might think that the market itself would provide adequate incentives for the pharma companies to produce effective medicines. Elixirs that actually cured the ailments they promised to cure would sell more than elixirs that were predicated on junk science. But the market mechanisms behind medical drugs were complicated by two factors that do not apply to most other consumer products. The first is the placebo effect. On average, human beings do tend to see improved health outcomes when they are told they are being given a useful medicine, even if the medicine they're taking is a sugar pill. How placebos actually work is still not entirely understood, but the effect is real. There is no equivalent placebo effect for, say, televisions or shoes. If you go into business selling fake televisions, 20 percent of your customers are not going to somehow imagine fake television shows when they get their purchase back to their living rooms. But a pharma company selling fake elixirs will reliably get positive outcomes from a meaningful portion of its customers.

The other reason market incentives fail with medicine is that human beings have their own internal pharmacies in the form of their immune systems. Most of the time when people get sick, they get better on their own—thanks to the brilliant defense system of leukocytes, phagocytes, and lymphocytes that recognizes and fights off threats or injuries and repairs damage. As long as your magic elixir didn't cause kidney failure, you could sell your concoction to consumers and most of the time they would indeed see results. Their strep throat would subside, or their fever would go down—not because they'd ingested some quack's miracle formula but because their immune system was quietly, invisibly doing its job. From the patient's point of view, however, the miracle formula deserved all the credit.

But the placebo effect and the immune system were no match for diethylene glycol. The deaths caused by Harold Watkins's elixir ended up triggering a kind of immune response from the government instead. Henry Wallace's testimony had revealed how powerless the FDA really was when it came to regulating pharmaceutical reform. Outraged citizens pressed for reform, and in 1938, Franklin Roosevelt signed the Food, Drug, and Cosmetic Act into law. For the first time, the FDA was empowered to investigate the safety of all drugs sold in the United States. At long last, the regulators could look beyond the raspberry flavoring to the more pressing question of whether the drug in question might kill you.

A YEAR BEFORE the Elixir Sulfanilamide crisis erupted, Eugene Geiling, the University of Chicago pharmacologist who would later identify the toxins in the elixir, received an inquiry from a precocious Canadian grad student named Frances Oldham, expressing interest in a position at Geiling's lab. Just twenty-one years old, Oldham had graduated from high school at fifteen, and had already completed a graduate degree in pharmacology at McGill. The letter and CV from the Canadian prodigy so impressed Geiling that he sent a response via Airmail Special Delivery. "If you can be in Chicago by March 1st," he wrote, "you may have the Research Assistantship for four months and then a scholarship to see you through a PhD. Please wire immediate decision."

There was just one catch. Geiling had addressed the letter to "Mr. Oldham." But Frances Oldham was, in fact, a woman—in an age when female biochemists were practically unheard of. "Geiling was very conservative and old-fashioned," Oldham later wrote, "and he really did not hold too much with women as scientists." She weighed sending a response back to Geiling noting the confusion. "Here my conscience tweaked me a bit," she recalled. "I knew that men were the preferred commodity in those days. Should I write and explain that Frances with an 'e' is female and with an 'i' is male?" Oldham ran the question by her McGill adviser, who dismissed her concerns. "Don't be ridiculous," he said. "Accept the job, sign your name, put Miss in brackets afterwards, and go!"

The decision proved to be a turning point for Oldham. "To this day," she wrote in her memoirs, "I do not know if my name had been Elizabeth or Mary Jane, whether I would have gotten that first big step up."[6]

One of her initial assignments was observing the rats during the animal tests of Elixir Sulfanilamide. The experience left an indelible impression on the twenty-two-year-old scientist: a belief that these kinds of mass tragedies—true betrayals of the Hippocratic oath— could be avoided with empirical lab analysis and the right regulatory oversight.

Decades later, Oldham would play a critical role in another milestone piece of legislation, this one also triggered by mass tragedy. In August 1960, Oldham—now known by her married name of Frances Oldham Kelsey—took a job at the FDA as one of only three medical reviewers, assessing the applications for new drugs. The FDA's oversight of the drug industry had expanded since the days of Elixir Sulfanilamide, but a number of significant limitations continued to hamstring the agency's ability to keep the drug supply safe. The FDA had only sixty days to approve or reject a new medicine; if the medical reviewers failed to make a determination during that time, the manufacturer was free to bring it to market. Most astonishingly, the manufacturer had no obligation to submit proof that the new drug actually *worked*. If the FDA was satisfied that a new drug wasn't dangerous, the agency would allow its pharma company to bring it to market. The manufacturers could stir together a random cocktail of ingredients and call it a cure for arthritis, and as long as it didn't contain any known toxins, they could sell it by the barrel to unwitting customers.

In a strange echo of her experiences as a young research assistant more than two decades earlier, Frances Oldham Kelsey found herself in the middle of an epic health crisis within a matter of weeks of starting her new job at the FDA. (She had a Zelig-like quality where medical disasters were concerned.) A few years before Kelsey moved to the FDA, a German company had begun selling a sleeping pill and antianxiety medicine with the trade name Contergan. It was later marketed as a treatment for morning sickness. The active ingredient in the medicine—an immunomodulatory drug called thalidomide—seemed to have miraculous powers: it made people sleepy and relaxed like other sedatives that had recently come on the market, but unlike those sedatives, tests suggested that it was

impossible to overdose on the drug. By 1960, it had been licensed
for use in over forty countries around the world.

That was when an application for the production and sale of
thalidomide—marketed in the United States as Kevadon—came
across the desk of Frances Oldham Kelsey.

Because thalidomide had been approved for use throughout
Europe, the American company that had licensed the drug,
Richardson-Merrell, had submitted a somewhat perfunctory new
drug application, or NDA. As medical reviewer, Kelsey's job was to
review the clinical trials and other supporting evidence that the company had submitted to demonstrate the drug's safety. In the case of
Kevadon, Richardson-Merrell had merely submitted testimonials
from doctors, not empirical studies. The pharmacologist working
for the FDA also had some questions about the way in which the
drug was absorbed that were not addressed in the submitted NDA.
Kelsey decided to declare the application incomplete, giving the
FDA another few months to review.

Shortly after issuing this ruling, Kelsey ran across an article in
The British Medical Journal documenting cases of neuritis—a kind
of nerve damage, potentially irreversible—associated with thalidomide use. The representative from Richardson-Merrell claimed to
have heard nothing about these reports, and after a trip to Europe
to investigate, informed Kelsey that the side effect "was not particularly serious and possibly was tied in with an inadequate diet."
The company soon adopted a new strategy with the FDA, empha-
sizing how much easier it was to overdose on other sleeping pills,
like barbiturates, that had been previously approved. "If Marilyn
Monroe had taken thalidomide," the company argued, "she would
still be alive."[7] But Kelsey was undeterred. The studies showing
nerve damage made her curious about the effect of the drug on a

growing fetus, given that many women were taking it as a treatment for morning sickness.

The hunch proved to be a tragically perceptive one. Unbeknownst to Kelsey, German obstetricians had already begun reporting an unusual surge in children born with severely malformed limbs, a condition known as phocomelia. Half of the newborns died. Once again, a frantic race to identify the culprit began. By the fall of 1961, with the Kevadon application still under review thanks to Kelsey's objections, European authorities had convincingly linked thalidomide to the wave of birth defects. In March 1962, Richardson-Merrell formally withdrew its application. More than ten thousand children were born around the world with phocomelia that had been caused by thalidomide, and an untold number died in utero. Only a handful of cases were reported in the United States. Americans had been spared the tragedy of thalidomide thanks to the discerning eye of Frances Oldham Kelsey and her colleagues at the FDA. In a Rose Garden ceremony, President Kennedy awarded her the President's Award for Distinguished Federal Civilian Service. "I thought that I was accepting the medal on behalf of a lot of different federal workers," she later wrote in her memoirs. "This was really a team effort."

We do not typically hear a lot about heroic bureaucrats, because the power of an effective bureaucracy like the FDA lies in part in the way its intelligence and expertise is distributed across thousands of people, each quietly doing his or her job reviewing the clinical records, interviewing the applicants, trying to understand the problem at hand with as much rigor as possible. That kind of system rarely produces iconic figureheads, like the CEOs or television stars or professional athletes who rise to prominence in other organizations. And because their trade does not naturally lend itself to

President John F. Kennedy
presents the President's Award for
Distinguished Federal Civilian Service
to Dr. Frances Kelsey, 1962
(WDC Photos / Alamy Stock Photo)

narratives of epic individual achievement, the value of that work has
long been underestimated by the general public. Calling someone
a "government regulator" is practically a slur in most mainstream
American politics today.

Yes, bureaucracies can stifle innovation. Yes, some regulations
can linger well past their sell-by dates. We need better mechanisms
to prune outdated codes. But where medicine is concerned, the
benefits of government oversight are vividly rendered in the sheer
scale of the lives saved once the so-called bureaucrats were empowered to actually investigate the safety of the drugs being sold to the
American people. Those benefits have real numbers behind them—
104 people who lost their lives to Elixir Sulfanilamide would have
survived had the FDA done the simplest animal tests on the drug;
and thousands of Americans might never have been born or might

have come into the world with terrible physiological disadvantages had Frances Oldham Kelsey arrived on the job sixty days too late.

Like the Elixir Sulfanilamide crisis before it, the thalidomide scandal immediately opened doors to new legislation that activists had been unsuccessfully promoting for years. Within a few months of thalidomide's being pulled off the market, Congress passed the landmark Kefauver-Harris Drug Amendments that radically extended the demands made on new drug applicants. The amendments introduced many changes to the regulatory code, but the most striking one was this: for the first time, drug companies would be required to supply proof of *efficacy*, not just safety. It wasn't enough for Big Pharma to offer evidence that they weren't poisoning their customers. Now, at long last, they would have to actually show proof that they were curing them.

The chronology here seems absurd on the face of it. How is it possible that we started asking the pharmaceutical companies for empirical success rates only a half century ago? But the truth is that the question of efficacy was a harder one to answer back when Frances Oldham first arrived at that University of Chicago lab. In 1937, the FDA couldn't have reasonably asked for proof of efficacy because the world of experimental medicine didn't have a standardized way of establishing successes or failures. But by the time Frances Oldham Kelsey showed up for her first day of work at the FDA in 1962, it did. Something fundamental changed in the quarter century that separated the two crises. Human beings had acquired a new superpower. It was not a superpower that looked impressive on the newsreels, like dividing the atom, or sending astronauts into space. It was a medical breakthrough, but not one that involved syringes or chemistry. It was closer to Farr's life tables: a breakthrough in the way we looked at data. The formal name for the innovation was randomized, controlled double-blind trials, usually referred to with the shorthand RCT. Of all the late arrivals of intellectual and technological history—the bicycles and the typewriters—the RCT may well be the most puzzling, and the most consequential.

THERE ARE FEW METHODOLOGICAL REVOLUTIONS in the history of science as significant as the invention of the RCT. (Only the seventeenth-century formulation of the scientific method itself—building hypotheses, testing them, refining them based on feedback from the test—looms larger.) Like those empirical methods that Francis Bacon and other proto-Enlightenment scientists developed, the RCT is a surprisingly simple technique—so simple, in fact, that it begs the question of why it took so long for people to discover it. The main ingredients of an RCT are visible in the term itself: randomized, double-blind, and controlled. Say you are testing a new drug that allegedly cures strep throat. First you find a large number of people who currently are suffering from the illness, and you randomly divide them into two groups. One group—known as the experimental group—will receive the medicine you are testing; the other will receive a placebo. The placebo group is the control: a kind of yardstick against which you can measure the effectiveness of the drug. The control group measures how long it takes for strep throat to be naturally cured by the body's immune system. If the medicine in question actually works, the group that has received the drug will get better faster than the control group. If there's no difference in outcome between the two—or if the experimental group starts dying of kidney failure—you know you have a problem with the drug you are testing. Crucially, in a true double-blind experiment, neither the people administering the experiment nor its participants know which subject is in which group. Withholding that knowledge prevents subtle forms of bias from creeping into the study. Once the data has been assembled, statistical analysis is performed to determine if one group was significantly better or worse off than the other group. Generally, the standard is demonstrating that the finding is less than 5 percent due to chance. In other words, if you ran the study one hundred times, more than ninety-five of those trials would show that the treatment produced positive results in the experimental group

Put all those elements together and you have a system for separating the quack cures from the real thing, one that avoids the many perils that had long bedeviled the science of medicine: anecdotal evidence, false positives, confirmation bias, and so on. When the FDA began demanding proof of efficacy from the drug manufacturers in 1962, they could make that demand because a system was now in place—in the form of the RCT—that could meaningfully supply that kind of proof.

The RCT emerged as a confluence of several distinct intellectual tributaries. As far back as 1747, the Scottish doctor James Lind had famously conducted a proto-RCT onboard the HMS *Salisbury*, in an attempt to determine a useful remedy for scurvy, which was at that point the leading cause of death in the nautical community. Lind's experiment took twelve sailors showing symptoms of the disease and divided them up into six pairs, giving each pair a different dietary supplement: cider, diluted sulfuric acid, vinegar, seawater, citrus, or a common purgative. While he did not include a proper control group that was given a placebo, he did attempt to keep all the other environmental factors the same for the subjects: giving them all the same diet (other than the supplements), and ensuring that they were exposed to the same living conditions onboard the ship. Lind's experiment correctly determined that the citrus supplement was the only treatment to have a positive effect in combating the disease.

In many respects, Lind's study was far from the modern form of the RCT. For starters, it lacked placebos and blinding, and there were simply not enough participants in the study to make it statistically meaningful. The importance of randomization would not become apparent until the early twentieth, when the British statistician R. A. Fisher began exploring the concept in the context of agricultural studies as a way of testing the effectiveness of treatments on distinct plots of land. "Randomization properly carried out," Fisher argued in his 1935 book, *The Design of Experiments*, "relieves the experimenter from the anxiety of considering and estimating the magnitude of the innumerable causes by which his data may be disturbed."[8]

Fisher's work on randomization and experiment design in the 1930s caught the eye of an epidemiologist and statistician named Austin Bradford Hill, who sensed in Fisher's method a technique that could prove hugely beneficial for medical studies. Hill would later echo Fisher's description of the powers of randomization, writing that the technique "ensures that neither our personal idiosyncrasies (our likes or dislikes consciously or unwittingly applied) nor our lack of balanced judgement has entered into the construction of the different treatment groups—the allocation has been outside our control and the groups are therefore unbiased."[9] Hill recognized that the key to successful experiment design was not just the researcher's ability to produce promising drugs to test but also to remove his or her influence over the results of the experiment, the subtle contaminations that so often distorted the data.

As a young man, Hill had contracted tuberculosis while serving as a pilot in the Mediterranean, and so it was somewhat fitting that the first landmark study that Hill oversaw was investigating a new treatment for tuberculosis, the experimental antibiotic streptomycin. When the results were published in *The British Medical Journal* in 1948, the title only alluded to the content of the study: "Streptomycin Treatment of Pulmonary Tuberculosis." But the real significance of the study lay in its form. It is now widely considered to be the first genuine RCT ever conducted. Antibiotics, as we will see in the next chapter, turned out to be the prime movers that finally transformed the world of medicine into a net positive force in terms of life expectancy. It is likely not a coincidence that the first true miracle drugs and the first true RCTs were developed within a few years of one another. The two developments complemented each other: the discovery of antibiotics finally gave the researchers a drug worth testing, and the RCTs gave them a quick and reliable way to separate the promising antibiotics from the duds.

Hill's randomized, controlled investigation into the efficacy of streptomycin was a milestone in the history of experiment design. Its indirect effect on health outcomes—thanks to the countless RCTs that would follow in its wake—would have earned him a place in the pantheon of medical history had he never published another paper. But Austin Bradford Hill was just getting started. His next study would have a direct impact on millions of lives across the planet.

. . .

AT SOME POINT during the chaos of World War II, as the Blitz was terrorizing London, public health officials in England began detecting an ominous signal in mortality reports compiled by the registrar general. While thousands were dying in bombing campaigns and on the front lines in Europe, another kind of killer was growing increasingly deadly across the population: lung cancer. The surge in deaths was truly alarming. By the end of the war, the Medical Research Council estimated that mortality from carcinoma of the lung had increased *fifteen*-fold from 1922. Cigarettes were one of the suspected causes, but many people pointed to other environ-. mental causes: the exhaust from automobiles, the use of tar in roadways, other forms of industrial pollution.

A few months before Austin Bradford Hill published his tuberculosis study, the Medical Research Council approached Hill and another noted epidemiologist named Richard Doll, asking the two men to investigate the lung cancer crisis. Today, of course, even grade-schoolers are aware of the connection between smoking and lung cancer—even if some of them grow up to ignore it—but in the late 1940s, the link was not at all clear. "I myself did not expect to find smoking was a major problem," Richard Doll would later recall. "If I'd had to bet money at that time, I would have put it on something to do with the roads and motorcars."

Hill and Doll devised a brilliant experiment to test the hypothesis that smoking might be connected to the surge in lung cancer cases. The structure was a kind of inverted version of a traditional drug trial. The experimental group was not given an experimental medicine, and there was no placebo. Instead, the experimental group was made up of people with existing cases of lung cancer. Hill and Doll approached twenty different London hospitals to find a statistically meaningful group of lung cancer patients. They then recruited two distinct control groups at each hospital: patients suffering from some other form of cancer, and patients without cancer at all. For each member of the "experimental" group—that is, the group with lung cancer—they tried to match with a control patient who was roughly the same age and economic class, and who lived in the same neighborhood or town. With those variables the same in each group, Hill and Doll ensured that some confounding factor wouldn't contaminate the results. Imagine, for instance, that the lung cancer surge turned out to be caused by the industrial soot in Lancashire factories. An experiment that didn't control for place of residence or economic status (factory worker versus sales clerk, say) wouldn't be able to detect that causal link. But by assembling an experimental group and a control group that were broadly similar to each other in terms of demographics, Hill and Doll could investigate whether there was a meaningful difference between the two groups in terms of smoking habits.

In the end, 709 people with lung cancer were interviewed about their smoking history, with the same number in the control group. Hill and Doll created multiple tables that explored those histories along different dimensions: average cigarettes smoked per day; total tobacco consumed over one's lifetime; age when the subject began smoking. Once the numbers had been crunched, the results were overwhelming. "Whichever measure of smoking is taken," Hill and Doll wrote, "the same result is obtained—namely, a significant and clear relationship between smoking and carcinoma of the lung."[10] At the end of the paper they eventually published, Hill and Doll made a rough attempt to evaluate the impact of heavy smoking on the probability of contracting lung cancer. By their estimate, a person who smoked more than a pack a day was fifty times more likely to develop lung cancer than a nonsmoker. The number was shocking at the time, but we now know it to have been a wild understatement of the risk. Heavy smokers are in fact closer to *five hundred* times more likely to develop lung cancer than nonsmokers.[11]

Despite the overwhelming evidence the study conveyed, and the rigor of its experimental design, the 1950 paper they published— *Smoking and Carcinoma of the Lung*—was initially dismissed by the medical establishment. Years later, Doll was asked why so many authorities ignored the obvious evidence that he and Hill had accumulated. "One of the problems we found in trying to convince the scientific community," he explained, "was that thinking at that time was dominated by the discovery of bacteria such as diphtheria, typhoid, and the tubercle, which had been the basis for the big advances in medicine in the last decades of the nineteenth century. When it came to drawing conclusions from an epidemiology study, scientists tended to use the rules that had been used to show that a particular germ was the cause of an infectious disease."[12] In a sense, the medical establishment had been blinded by its own success identifying the causes of other diseases. While an overwhelming number of lung cancer patients had turned out to be heavy smokers, there were still a number of nonsmokers who had suffered from the disease. Using the old paradigm, those nonsmokers were like finding a cholera patient who had never ingested the *Vibrio cholerae* bacterium. "But, of course, nobody was saying [smoking] was *the* cause; what we were saying is that it is *a* cause," Doll explained. "People didn't realize that these chronic diseases could have multiple causes."

Undeterred, Hill and Doll set out to conduct another experiment, approaching the smoking question from a different angle.

They decided to see if they could *predict* cases of lung cancer by analyzing people's cigarette use and health outcomes over many years. This time they used physicians themselves as the subjects, sending out questionnaires to more than fifty thousand doctors in the United Kingdom, interviewing them about their own smoking habits and then tracking their health over time. "We planned to do the study for five years," Doll later recalled. "But within two and a half years, we already had 37 deaths from lung cancer and none in nonsmokers." They published their results early, in 1954, in what is now considered a watershed moment in the scientific establishment's understanding of the causal link between smoking and cancer.

In that 1954 paper, the experiment design proved to be less important than the unusual choice of subjects. Hill and Doll had originally decided to interview physicians because it was easier to follow up with them to track their health and cigarette use over the ensuing years. But the decision proved to have additional benefits. "It turned out to have been very fortunate to have chosen doctors, from a number of points of view," Doll noted. "One was that the medical profession in this country became convinced of the findings quicker than anywhere else. They said, 'Goodness! Smoking kills doctors, it must be very serious.'"

Exactly ten years after the publication of Hill and Doll's second investigation into the link between cancer and smoking, surgeon general of the United States Luther Terry, a physician, famously issued his *Report on the Health Consequences of Smoking*, which officially declared that cigarettes posed a significant health threat. (After nervously puffing on a cigarette on the way to the announcement, Terry was asked during the press conference whether he himself was a smoker. "No," he replied. When asked how long it had been since he had quit, he replied, "Twenty minutes.") Subsequent studies modeled after Hill and Doll's pioneering work identified other health threats posed by smoking, including cardiovascular disease, now the number one killer in the United States. Government regulators around the world added warning labels to tobacco products; advertising restrictions were established; cigarettes were heavily taxed. When Hill and Doll interviewed their first patients in the London hospitals, more than 50 percent of the UK population were active smokers. Today the number is just 16 percent. Quitting smoking before the age of thirty-five is now estimated to extend your life expectancy by as much as nine years.

The partnership between RCT design and government regulation—the experiments revealing threats that governments then outlaw or restrict—led to a quiet but profound revolution in the health of millions of people around the world. Compounds used in the production of dyes and rubber were found to produce bladder cancer and were eliminated; the skin cancer generated by the exposure of road workers to tar was greatly reduced; asbestos was outlawed after studies linked it to the rare and deadly cancer mesothelioma. It was a revolution jump-started not by spectacular technological breakthroughs or protesters in the streets but instead by different kinds of agents: artful experiment designers, government regulators. It was a revolution in the kind of questions we asked, and the formal way we set about to answer them. Is this new elixir safe? Does it actually cure people? Are cigarettes dangerous? And how can we know for sure?

6

THE MOLD THAT CHANGED THE WORLD

ANTIBIOTICS

Anyone who has had a passing interest in the history of science and medicine has probably come across the fabled account of the discovery of the first true antibiotic, penicillin. It's a story that has become almost as familiar as the one about Newton's apple and the theory of gravity, in part because it shares the same structure of a fortuitous accident and a sudden stroke of insight. On a fateful day in September 1928, the Scottish scientist Alexander Fleming accidentally leaves a petri dish containing bacterium *Staphylococcus* exposed to the elements next to an open window and then departs for a two-week vacation. When he returns to his lab on September 28, he discovers that a blue-green mold has contaminated the *Staphylococci* culture. Before he can dispose of it, Fleming notices something strange: the mold appears to have inhibited the growth of the bacteria. His curiosity piqued, Fleming examines the culture plate more closely, and observes that the mold seems to be releasing some kind of substance that triggers lysis of the bacteria— breaking down their cell membranes, effectively destroying them. It's another holy grail: a bacteria killer. Fleming calls it penicillin. Seventeen years later, after the true magnitude of his discovery has become apparent, he is awarded the Nobel Prize in Medicine.

The Fleming story has traveled so widely in part because it has served as a justification for anyone who keeps a messy desk at work. If Fleming had been just a little tidier, he would almost certainly never have received that Nobel. (Indeed, there is a long tradition of generative clutter in the history of innovation: X-rays were discovered thanks to an equally disorganized work environment.) But like so many stories of genuine breakthroughs, the tale of the petri dish and the open window massively compresses the real narrative of how penicillin—and the antibiotics that quickly followed in its wake— came to transform the world. The triumph of penicillin is actually one of the great stories of international multidisciplinary collaboration. It is a story of a network, not an eccentric genius.

Fleming was part of that network, but only a part. He seemed to have not entirely grasped the true potential of what he had stumbled upon. He failed to set up the most basic of experimental trials to test its efficacy at killing *Staphylococci* outside the petri dish. "All Fleming had to do to demonstrate the curative effect of penicillin was to inject .5 ml of his culture fluid in to a 20 g mouse infected with a few streptococci or pneumococci," a contemporary noted. "He did not perform this obvious experiment for the simple reason that he did not think of it."[1]

This oversight was genuinely shocking, given how high the stakes were. Humans had been locked in a life-or-death struggle with bacterial diseases since at least the dawn of civilization. Skeletons excavated from Egyptian gravesites dating back six thousand years show signs of the deformities introduced by spinal tuberculosis. Hippocrates treated patients who had clearly been infected with the tuberculosis bacterium. For much of the nineteenth century, it was responsible for a quarter of all deaths. It may be, in the long view, the deadliest killer of all the infectious diseases. Bacterial infections caused by simple scrapes and cuts—or medical procedures—were also major killers. Some estimates suggest that two thirds of the deaths in the American Civil War were the result of sepsis and other infections acquired in military hospitals. The threat of infection was one major reason medical interventions had such a poor track record extending life expectancy well into the twentieth century. Even if the doctors did have the technical ability to save your life, they might inadvertently kill you through bacterial infections.

Those were the extraordinary stakes that surrounded Fleming's discovery of penicillin. A drug that might finally be able to make a direct assault on this ancient nemesis could usher in a true revolution in medicine. And as Fleming sat on his discovery through the 1930s, the stakes only grew higher with the first stirrings of what would become World War II. In the end, it was the carnage of a global military conflict that turned Fleming's discovery into a true lifesaver.

THERE ARE LITERALLY THOUSANDS of stories to be told about the impact of antibiotics on the Second World War, one for each of the many lives saved by penicillin, and for the ones lost because the miracle drug wasn't available. But consider this one story as a representative sample: On May 27, 1942, the Nazi official Reinhard Heydrich was being driven through the suburbs of Prague in a Mercedes convertible, on his way to meet Hitler in Berlin. (Heydrich had been the principal organizer behind the Kristallnacht attacks, among other atrocities.) A team of Czech assassins trained by the British lay in wait at a hairpin turn in the road. As Heydrich's car slowed for the turn, one of the assassins pulled out a machine gun, but it jammed and failed to fire. The other tossed a grenade at the Mercedes that landed outside the rear of the vehicle but did some damage nonetheless. At first it seemed like a lucky break for the Nazis, given how vulnerable a target Heydrich had been. He was wounded but not fatally. After surgery to remove his spleen, doctors were optimistic that he would make a full recovery. But part of Heydrich's wounds involved splinters and horsehair from the seats of the Mercedes. Some microscopic organism entered his bloodstream through those seemingly minor wounds and began replicating. Within hours of the doctors' upbeat prognosis, the patient had developed blood poisoning.[2]

Heydrich died June 4, just a week after the attack. He had survived the explosive violence of machine guns and grenades. What killed him instead was an invisible threat: the bacteria that infected his wounds.

As it happens, Heydrich died at almost the exact moment that British and American scientists—supported by the US military—were for the first time producing enough stable penicillin to cure an infection like the one that had taken Heydrich's life. By the last years of the war, the Allies had penicillin in significant quantities, while the Axis powers never developed it. That gave the Allies a subtle but material advantage. The atom bomb might have ended the war in Asia, but you can make a convincing case that penicillin played a key role in securing victory in Europe. It was a defensive achievement: in part, the Allies won the war not by figuring out how to kill more of the enemy but rather by figuring out a new way to keep their soldiers from dying. It was a battle fought in hospitals, not on the front lines. But it was an achievement that mattered nonetheless. So how did it end up happening?

Part of that explanation certainly involves Fleming himself. While it is true that he failed to act in a meaningful way on his discovery, the fact that it was Fleming who made the original discovery of penicillin was not just a matter of a happy accident. He was precisely the sort of intellect who sought out interesting developments in chaotic environments. He was an avid game player, at both work and leisure. Whatever amusement he happened to be pursuing—golf or snooker or cards—he was constantly inventing new rules on the fly, sometimes midgame. When asked to describe his work, he would often describe it with a seemingly self-deprecating "I play with microbes."[3] But he meant it seriously. A mind less drawn to the surprising combinations that all play elicits would have taken one look at that moldy petri dish and dismissed it as garbage, a spoiled experiment. Fleming assumed it was interesting. That is often how new ideas come into the world: someone perceives a signal where others would instinctively perceive noise.

Fleming's playful relationship to his research was evident early in his career. As a student at St. Mary's Hospital Medical School in London, he created elaborate paintings using bacteria as the pigments, a technique that was predicated on Fleming's knowledge of the different colors expressed by each bacterium as it grew. Those microbial works of art may sound frivolous, but during that time frame—the first decade of the twentieth century—exploring the connection between bacteria and color was actually an incredibly fertile ground for scientific research, one that would eventually provide a crucial foundation for the antibiotics revolution. These discoveries, too, emerged from a seemingly unrelated field: fashion.

As late as the 1870s, the most advanced chemical companies in the world did most of their business by manufacturing dyes. "Dyes were by far the largest and most lucrative chemical process yet known," the medical historian William Rosen notes, "and enormously more profitable than, say, medicine."[4] Vegetable-based dyes had been adding color to fabrics for thousands of years, but advances in chemistry in the nineteenth century had opened up a new, tantalizing possibility: the creation of color that could stain fabric using synthetic materials. Because those new dyes could be produced at industrial scale, they quickly attracted the attention of entrepreneurs looking to capitalize on the new production techniques. Many of the companies formed during this period were based in Germany, including the conglomerate that eventually became known as IG Farben. That company would go on to generate a wide range of chemical breakthroughs—including polyurethane and, most notoriously, the poison Zyklon B used in the Nazi gas chambers—before being broken up after World War II. But its roots were evident in its name: *farbe* is the German word for color, and the verb *färben* means "to dye."

The flurry of interest in synthetic dyes led a whole generation of researchers to explore innovations in tissue staining, culminating in the work of Paul Ehrlich, who developed a series of techniques that could add color to individual cells based on their identity, making it possible to distinguish between different kinds of blood cells. Eventually these staining techniques were applied to distinguish between what are called gram-positive and gram-negative bacteria, a distinction that would become crucial to the development of antibiotics in the 1940s.

Some lucky accidents happen in the lab when a petri dish happens to get contaminated; other lucky accidents happen on a different scale, when one field of research unwittingly provides tools that can be used in a completely different field. In part we developed the ability to perceive otherwise invisible bacteria because scientists like Robert Koch used new microscopes in experiments that were specifically designed to explore the microbial world. But we also developed these new powers because there was money to be made selling people brightly colored clothing.

THERE IS ANOTHER ELEMENT to Alexander Fleming's role in the antibiotics revolution worth mentioning: he was working as part of the British medical establishment in the 1920s and 1930s, surrounded by some of the brightest minds in medical research during that period. Had he made his penicillin discovery Mendel-style, off in a monastery somewhere, it might well have gone nowhere, given Fleming's strange unwillingness to follow up with a rigorous test of its utility. But Fleming was part of a wider network, which meant that his work was likely to attract the attention of other researchers, with other kinds of skills. For penicillin to graduate from a brilliant accident to a true miracle drug, three things needed to happen: someone had to determine whether it actually worked as a medicine; someone had to figure out how to produce it at scale. And then a market had to develop to support that large-scale production.

All three of those key pieces came together in a remarkably short time, roughly between 1939 and 1942, a period of staggering chaos in global politics. In the late 1930s, two Oxford scientists, the Australian Howard Florey and the German-Jewish refugee Ernst Boris Chain, stumbled across a long-neglected paper that Fleming had published in 1929 on his penicillin discovery. Florey was the head of the Sir William Dunn School of Pathology at Oxford, an institute that had been founded just a few decades earlier to study pathogens and their effect on the humane immune system. Florey saw potential in the mysterious mold but thought it might be too difficult to reproduce the compound in a stable enough form to be used as a medicine. Chain, however, saw the instability as a challenge. But before they could work on stabilizing the drug, much less test it on animals, they had to figure out a way to produce sufficient quantities of the mold to do lab experiments with it. Fortunately for Florey and Chain, a junior member of the Dunn School team, Norman Heatley, was a brilliant laboratory engineer and a true polymath, trained in biology and biochemistry but also, in the words of one of Florey's biographers, "the technical skills of optics, glass and metalworking, plumbing, carpentry, and as much electrical work as was needed. And he could improvise—making use of the most unlikely bits of laboratory or household equipment to do the job with the least possible waste of time."[5]

After a furious period of trial and error, Heatley designed a bizarre contraption hacked together from a collection of lab equipment and motley spare parts, including a recycled doorbell, baling wire, cookie tins, bedpans, and a sewing needle employed to create precise holes in hot glass. Rosen describes the Rube Goldberg device Heatley engineered:

> Three bottles—of broth, ether, and acid—are held upside
> down in a frame, until the glass ball stopper in the bottle

containing broth is moved aside; liquid flows into a glass coil surrounded by ice. Once cooled, the acidified liquid combines with acid from bottle number three and is jet-sprayed in droplets that arrive in one of six parallel separation tubes. Meanwhile the stopper on bottle number two, containing ether, is moved aside, releasing ether into the bottom of the whole arrangement. The filtrate in the separation tube is sprayed into a tube of ether rising in a four-foot-long tube. As penicillin has a chemical affinity for the ether, it transfers into that tube, leaving the remaining components of the original broth behind, to be drained out. Then, the penicillin-plus-ether (later acetate) solution is introduced into another tube, with slightly alkaline water. The penicillin-plus-water mixture—about 20 percent of the volume of the filtered broth that started the whole rigmarole—was drawn off.[6]

Heatley's device made Alexander Fleming's workspace look orderly by comparison, but it worked: the contraption could transform twelve liters of moldy "broth" into two liters of functional penicillin in just an hour.

On May 25, 1940, Florey performed the first real test of penicillin's effectiveness. He deliberately infected eight mice with the bacteria responsible for strep throat, and other even more debilitating disorders. He then gave four of them penicillin—in differing doses—and gave the remaining four nothing. It was not a proper RCT, but the results were striking enough that Florey knew he was onto something. All four controls died. The ones that had been given penicillin all lived.

After further experiments and lab engineering had allowed the

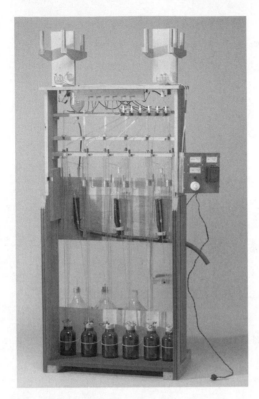

Replica of the apparatus for the continuous extraction and purification of penicillin, re-created by Dr. Norman Heatley for the UK Science Museum, 1986

Dunn School team to produce even purer versions of the drug, they decided to test penicillin on an actual human subject. Florey dispatched a young researcher to find a patient willing—or desperate enough—to participate in the experimental trial. Charles Fletcher knew exactly where to look. "Every hospital then had a septic ward," Fletcher later wrote. The primary treatment for acute infections was simply the application of bandages. "There was nothing else," Fletcher observed. "About half the patients who came to these wards died."[7]

Fletcher quickly identified an ideal test case in a nearby Oxford hospital, a patient whose condition is a reminder today of the kind of grotesque infections that used to originate in the smallest of scrapes in the era before antibiotics. A policeman named Albert Alexander had scratched his face on a rose thorn while gardening. It had seemed like a minor annoyance at the time, but below the surface of the wound, *Staphylococcus* bacteria that originally lived in the garden soil began replicating. By February, the infection had spread throughout his body. He had lost his left eye to the bacteria. Fletcher later wrote, "He was in great pain and desperately and pathetically ill." The night after Heatley visited Alexander in the hospital, he wrote in his diary, "He was oozing pus everywhere."[8] Without a miracle drug to fight the infection, Alexander was clearly going to be dead in a matter of weeks, if not days.

Florey and his Dunn School colleagues decided that Alexander fit the bill for their drug trial. On February 12, 1941, Albert Alexander was administered 200 mg of penicillin. Every three hours after the initial dose, he received an additional 100 mg. A contemporary hospital encountering a patient in such critical condition would have slammed him with more than twice that dosage. But Florey was just guessing at the appropriate levels for the drug; this was the first human to be treated, after all.[9] No one knew what amount would be useful and what amount might be lethal.

Florey's educated guess turned out to be right. Within hours, Alexander began to heal. It was like watching a reverse horror movie: a man whose body had been visibly disintegrating suddenly switched directions. His temperature settled back to a normal range; for the first time in days he could see through his remaining eye. The pus that had been dripping from his scalp had entirely disappeared.

As they watched Alexander's condition improve, Florey and his Dunn School colleagues recognized they were seeing something genuinely new. "Chain was dancing with excitement," Fletcher would write of the momentous day. "Florey [was] reserved and quiet but nonetheless intensely thrilled by this remarkable clinical story." For the first time in the long coevolutionary dance between bacteria and humans, the humans had devised a reliable technique to kill the bacteria, not by washing hands or purifying water systems, but by engineering a new compound that could be ingested by infected people and distributed through their bloodstreams to attack the killer microbes. Vaccines had fought off pathogens by triggering our immune systems. Public health had done it by building external immune systems. Penicillin was a new trick: manufacturing a compound that had its own pathogen-killing powers.

Yet for all the genius of the Dunn School team, they had not yet solved the scale problem. In fact, they had such limited supplies of penicillin that they took to recycling the compound that had been excreted in Alexander's urine. After two weeks of treatment, they ran out of the medicine. Alexander's condition immediately worsened, and on March 15 the policeman died—killed by a rose thorn scratch. His remarkable, if temporary, recovery had made it clear that penicillin could cure people of deadly bacterial infections. Left unanswered was whether you could produce enough of the stuff to make a difference.

TO SOLVE THE SCALE PROBLEM, Howard Florey turned to the Americans. He wrote Warren Weaver, the visionary head of the Rockefeller Foundation, explaining the Dunn School's promising new medicine. Weaver recognized the significance of the finding and arranged to have the penicillin and Florey and Heatley brought over to the United States to pursue their research there, far from Blitz-rattled England. In a scene straight out of *Casablanca*, Florey and Heatley took the Pan Am Clipper from Lisbon on July 1, carrying a locked briefcase containing a significant portion of the world's supply of penicillin.

In America, the team was quickly set up with a lab at the US Department of Agriculture's Northern Regional Research Laboratory in Peoria, Illinois. Almost immediately, the project attracted the support of the US military, who were understandably eager to find a miracle drug that would protect the troops from infections that had killed so many soldiers in past conflicts. Before long, several American drug companies—including Merck and Pfizer—were enlisted as part of the project as well, given their expertise at mass production. For Florey and Heatley, Peoria might have seemed like a remote outpost for a project that had up to that point thrived in the dense intellectual networks of London and Oxford, but the facility turned out to be an ideal setting. The agricultural scientists had extensive experience with molds and other soil-based organisms. And the heartland location had one meaningful advantage: its proximity to corn. The researchers at the DOA lab had been studying the fermentation powers of corn steep liquor, a waste by-product expelled in the creation of cornstarch. The mold turned out to thrive in vats of corn steep liquor.

The team decided to tackle the scale problem from two angles. They continued with the lab engineering approach that Heatley had so brilliantly developed back in Oxford, building new contraptions to maximize the yield from an original supply of the mold, now suffused with corn steep liquor. But they also suspected that there might be other strains of penicillin out in the wild that would be more amenable to rapid growth. The agronomists in Iowa knew that ordinary soil was teeming with both bacteria, like the

Staphylococci that had killed Albert Alexander, and with organisms—like Fleming's original mold—that had evolved defenses to keep those bacterial threats at bay. They could waste months trying to breed mold more effectively in Heatley's contraptions, while an organism that was much more amenable to mass production might be lying in the dirt somewhere.

And that is how the United States government came to launch one of the greatest needle-in-a-haystack operations in the history of the world, only in this case the needle was a mold that might well have been invisible to the naked eye, and the haystack was anywhere on the planet that had live soil. While Allied soldiers fought the iconic battles of World War II, dozens of soldiers quietly pursued a separate mission all around the world, a mission that on the face of it seemed closer to kindergarten recess than military action—literally digging in the dirt, collecting soil samples to be shipped back to the American labs for investigation. One of those expeditions brought back an organism that would become the basis for streptomycin, now one of the most widely used antibiotics in the world, and the basis for Austin Bradford Hill's pioneering RCT in 1948. In the years immediately after the end of the war, drug companies such as Pfizer would go on to conduct massive exploratory missions seeking out soil samples from every corner of the planet. According to a Pfizer chemist, the company "got soil samples from cemeteries; we had balloons up in the air [that] collected soil samples that were windborne; we got soil from the bottoms of mine shafts . . . from the bottom of the ocean."[10] A staggering 135,000 distinct samples were collected.

In Peoria, the team conducted its own search for alternative strains of penicillin. During the summer months of 1942, shoppers in local grocery stores began to notice a strange presence in the fresh produce aisles: a young woman intently examining the fruit on display, picking out and purchasing the ones with visible rot. She must have seemed to be an eccentric customer to the grocers and checkout clerks, but in reality she was on a top secret mission, integral to the life or death of millions of Allied troops fighting the war. Her name was Mary Hunt, and she was a bacteriologist from the Peoria lab, assigned the task of locating promising molds that might replace the existing strains that were being used. (Her unusual shopping habits ultimately earned her the nickname Moldy Mary.) One of Hunt's molds—growing in a particularly unappetizing cantaloupe—turned out to be far more productive than the original strains that Heatley and the Dunn School team had tested.[11] Because of Alexander Fleming's original discovery, the penicillin narrative is commonly presented as a case study of someone stumbling across a new idea by chance, and being receptive enough to see something intriguing in that new combination. But the triumph of penicillin is also a story of deliberate search, not just accidental discovery. Mary Hunt was searching through those rotten cantaloupes because she thought they might harbor a killer mold, and because an entire team of Allied scientists had become convinced that such a discovery could be useful in the war effort.

They were right, all of them. Nearly every strain of penicillin in use today descends from the bacterial colony Hunt found in that cantaloupe.

Aided by the advanced production techniques of the drug companies, the United States was soon mass-producing a stable penicillin in quantities sufficient to be distributed to military hospitals around the world. When the Allied troops landed on the Normandy beaches on June 6, 1944, they were carrying penicillin along with their weapons.

. . .

AS IS SO OFTEN the case with significant innovations, we cannot say with any certainty exactly when penicillin was invented. The answer to that question is a range, not a point on a time line. All we can say is that the miracle drug of antibiotics did not exist in any real sense before 1928, and by the middle of 1944 it was a material force in the world, saving thousands of lives a week, and giving the Allied countries a quiet but critical advantage over the Axis powers. An absentminded professor with a messy lab does indeed mark the beginning of that revolution, but there are many such stories in the history of innovation, even in medical innovation. What made the penicillin revolution so different was how quickly the insight from the cluttered laboratory was able to make it to mass production, thanks largely to the scaling powers of the United States military and the private drug companies. The compound itself was part of what had to be discovered and refined to bring penicillin to the world, but it was just as important that we invented new paths of sharing and amplifying the discovery: from Fleming's lab to the Dunn School to Peoria and to the Normandy beaches.

What was the impact of that revolution, all told? The discoveries of penicillin and its successor antibiotics (almost all of which were developed in the two decades that followed Florey and Heatley's first successful test in 1942) directly saved hundreds of millions, if not billions, of lives around the world. Before Fleming left that petri dish exposed to the elements, tuberculosis was the third most common cause of death in the United States; today it is not even in the top fifty. The magical power of antibiotics to ward off infection also opened the door to new treatments: radical surgical procedures like organ transplants that were severely vulnerable to life-threatening infections became far safer, allowing them to enter the mainstream of medical practice. The antibiotics revolution also marked a watershed moment in the history of medicine. Thanks to these miracle drugs, medicine finally broke free of the bleak constraints of the McKeown thesis. While a handful of new medicines before penicillin had improved health outcomes—Paul Ehrlich's original "magic bullet" treatment for syphilis, Salvarsan, along with insulin injections for diabetics and the sulfa drugs of the 1930s—antibiotics offered an unprecedented line of defense against disease and infection. Starting in the postwar years, human life expectancy was not just being extended by public health institutions and pasteurized milk, but also by pills that finally offered something more useful than a mere placebo effect. Hospitals are no longer places where we go to die, offering nothing but bandages and cold comfort. Routine surgeries rarely result in life-threatening infections. Over the subsequent decades, antibiotics were joined by other new forms of treatment: the statins and ACE inhibitors used to treat heart disease; a new regime of immunotherapies that hold the promise of curing certain forms of cancer for good. The model of serendipitous drug discovery that defined the search for the first generation of antibiotics—all those molds extracted from soil samples all around the world—has been increasingly replaced by a new approach, sometimes called "rational drug design," in which new therapeutic compounds are designed using computers, based on our knowledge of the molecular receptors on the surface of a virus or other disease agents. (The AIDS cocktail that has saved so many millions of lives over the past two decades was one of the first triumphs of the rational design approach.) Quack cures remain on the market, but most of the items offered for sale by reputable drug companies actually perform as advertised. It took longer than we might have naturally expected, but today's medical healers, armed with penicillin and its many descendants, have finally developed the ability to cure diseases, not just prevent them.

The discovery and amplification of penicillin is a reminder that we cross disciplines for the same reason we cross strains of wheat: they become more resilient, more fertile when we do. What did it take to get from 1928 to 1942 in our penicillin time line? It took a chaotic workspace, soil scientists, a grocery store, a vat of corn steep liquor, and an entire military apparatus. It took chemists and industrial engineers. And all of those actors were relying on insights and technology that had originated with the lens makers and the fabric industry and the farmers of the nineteenth century. When you perceive the whole network this way, it almost seems like one of Norman Heatley's eccentric contraptions, a chain of unlikely bedfellows strung together. It is not quite as clean a narrative as the classic cliché of a genius at the microscope, but it is a more accurate account of how something as transformative as penicillin becomes a part of everyday life.

FOR UNDERSTANDABLE REASONS, the history of innovation—medical or otherwise—tends to be organized around momentous, singular breakthroughs: penicillin or the smallpox vaccine. But it can sometimes be as instructive to investigate why a specific breakthrough *didn't* come into being in a given society. The question of why the Nazis were not able to develop an atomic bomb—and the potential consequences if they had been able to—has been pondered many times over the years. But just as interesting is the question of why they were unable to develop penicillin.

One factor may have been the German investment in the class of drugs known as the sulfonamides, the early predecessor to antibiotics that had killed so many Americans in the 1937 incident. The sulfa drugs had been originally developed in the early 1930s at the German chemical and pharmaceutical conglomerate IG Farben. Sulfa drugs could combat bacterial infection—Allied troops had carried sulfa packets before penicillin was introduced—but bacteria easily developed resistance to them, and the drugs themselves could be toxic. The fact that Germany already had significant commitments to the mass production of sulfonamides—perhaps aided by a nationalistic pride in their discovery—may have made them less likely to investigate other alternatives. As with the atomic bomb, the brain drain of scientists, many of them Jewish, who had fled in the buildup to the war, gave the Allies an additional advantage, most obviously in the form of Ernst Boris Chain. Many of the chemists who remained were focused more on developing lethal gases to carry out the Final Solution than they were on life-saving medicines.

One additional factor was undoubtedly the secrecy that surrounded the project on the American side. While Fleming's original work—along with some of the Oxford breakthroughs—were a matter of public record, by the time the team began making significant progress in Peoria, the US government had recognized the strategic advantage that the miracle drug might give them against the Nazis. Twelve days after Pearl Harbor, President Roosevelt established an emergency wartime agency known as the Office of Censorship, assigned the task of monitoring—and where necessary, impeding—the flow of information to the country's enemies. In subsequent histories, the office's most celebrated activities involved its top secret support of the Manhattan Project. But the day after

Roosevelt created the office, the team in Peoria was informed that "any information relevant to the production and use [of penicillin] should be severely restricted."[12]

The Nazi regime did make some attempts to produce the drug at scale. A small team of scientists in the Hoechst dye works began investigating the drug in 1942, but the project lagged far behind the developments in Peoria. Hoechst was not able to shift from small batch laboratory production to factory production until late 1944. Hitler and his deputies appear to have recognized the potential benefit of the drug; a cable from Berlin to Hoechst in March 1945 demanded an accounting of how many tons of penicillin they could produce each day. Even at that stage, the request was delusional; the Hoechst chemical plants were nowhere near that level of productive capacity. And just days after the cable arrived, the Hoechst dye works was seized by Allied soldiers, putting an end to the Nazi's belated quest for the miracle drug.

There is a curious footnote to the story of penicillin and World War II. On July 20, 1944, a little more than a month after the Allied forces had landed at Normandy, a bomb planted in a conference room at the Wolf's Lair military headquarters nearly assassinated Hitler. During the blast, Hitler suffered cuts, abrasions, and burns; many of his wounds contained wooden splinters from the conference room table that had protected him from the full force of the blast. Recognizing the risk of infection that had killed Reinhard Heydrich two years before in Prague, Hitler's doctor, Theodor Morell, treated his wounds with a mysterious powder. In his journals, Morell referred to Hitler as "Patient A"; his notes from the night of July 20 read as follows:

"Patient A: eye drops administered, conjunctivitis in right eye.

One fifteen P.M. Pulse 72. Eight P.M. Pulse 100, regular, strong, blood pressure 165–170. Treated injuries with penicillin powder."[13]

Where did Morell get this penicillin? The Hoechst labs had barely begun even small-scale production in July of 1944, and it was unclear whether the drugs they were producing at that stage were even effective. But Morell had access to another supply of the miracle drug, a few ampules that had been discovered on captured American soldiers and passed on to Morell by a German surgeon. After the July 20 bombing, another doctor implored Morell to use some of the stolen antibiotics to treat another Nazi who had been horribly injured in the blast. Morell refused, presumably reserving his supply of high-grade penicillin for the Führer. One can only speculate on the course of events that would have followed had Hitler developed the same sort of fatal infection that had taken Heydrich's life. Almost certainly the war would have ended months earlier than it did. But whatever the implications, Dr. Morell's journal entry does suggest an ironic twist to the story of the international network that brought penicillin to the masses. Fleming, Florey, Chain, Heatley, Mary Hunt—they all played an integral role in helping the Allies triumph over Nazi Germany. They also may have saved Hitler's life.

7

EGG DROPS AND ROCKET SLEDS

AUTOMOBILE AND INDUSTRIAL SAFETY

O n August 31, 1869, the Irish scientist and aristocrat Mary Ward went for a drive with her husband and cousin through the back roads of County Offaly, in the Irish midlands. They were riding in an experimental steam-powered vehicle, a predecessor of the automobile. (Her cousin's sons had built the prototype steam car themselves.) It was typical of Mary Ward to be swimming in such adventurous waters. Despite the gender conventions of the day, she had carved out a career for herself as an astronomer and a science writer; she was particularly adept at using the newly crafted microscopes that were appearing during this period, powered by those new glass lenses that were about to reveal an entire hidden ecosystem of microbes. She was also an accomplished artist. Ward published several books featuring elaborate illustrations of what she had uncovered in her own microscopic explorations.

Mary Ward had followed a noteworthy path in the years preceding that August day in 1869. Had she lived to a ripe old age and died in her sleep, she would have been remembered for her achievements as a scientist—and as a popularizer of science—in an age when such achievements were hard won when the scientist in question was a woman. Instead, she is mostly remembered for how her exemplary life ended.

The modern eye would not be impressed by the cumbersome steam car that Ward and her fellow riders were riding in. The technology was known as road locomotion in the jargon of the day. It had the centaur-like look of a miniature train attached to the back of a (horseless) carriage. Driver and passengers sat up front and controlled the wheels with a lever. But as awkward as the device looks to us now, there was an understandable logic to it, given the technology that had preceded it. Steam-powered locomotion had revolutionized travel by rail. The next frontier was sure to be the existing road system. And so a whole generation of engineers slapped miniaturized steam engines onto drive trains and started whirling around the countryside.

Whirling might have been an overstatement. The maximum speed of these vehicles was somewhere in the range of ten miles per hour, and most local ordinances that had managed to keep up with the technology forbade drivers to exceed five miles per hour. But the road locomotives were heavy enough to be a menace even at those low speeds. Subsequent testimony estimated that the vehicle carrying Mary Ward was traveling less than four miles per hour on that August day in 1869. But when her party was rounding a sharp corner near a church in the town of Parsonstown, a sudden jolt threw Ward from the carriage. The rear wheels of the vehicle crushed her neck. After her husband and fellow passengers leaped from the vehicle, they found her bleeding from her ears, mouth, and nose, and in convulsions. Within minutes she was dead.

The next day the local paper ran a mournful account of her death. "The utmost gloom pervades the town," it read, "and on every hand sympathy is expressed with the husband and family of the accomplished and talented lady who has been so prematurely hurried into eternity."[1] Short notices on the accident appeared in papers across England and Ireland, with headlines like FATAL AC-CIDENT TO A LADY and FEARFUL DEATH OF A LADY. The readers of those news items had no idea that Mary Ward's accident would turn out to be the first in an unimaginably long list of fatalities with the same underlying culprit. The cause of death was ultimately declared by the coroner to be a broken neck, and a jury subsequently de-clared the death an accident. But attributing her death to a broken neck was like attributing a cholera death to dehydration. It was technically true, but the real villain was elsewhere. Mary Ward was killed by a machine. She is believed to be the first person to die in a motor vehicle accident.

Given the existing categories utilized by mortality reports dur-ing that period, Mary Ward's death was likely included in the ac-cident category. But soon enough, the public health officials had to introduce a new, more specific category into the taxonomy: auto-mobile deaths. Just as medicine was finally maturing into a genu-inely life-saving practice in the middle of the twentieth century, a new self-imposed threat emerged to shorten our lives. Back when

Henry Ford was inventing the Model T, tuberculosis was the third leading cause of death in the United States. But by the time antibiotics reached the masses in the early 1950s, it had been replaced on the list by the entirely man-made menace of the automobile.

MOST OF THE STORY of our doubled life expectancy comes from triumphing over threats that we had faced for millennia: killer viruses, bacterial infections, hunger. But starting in the nineteenth century a genuinely new kind of threat emerged, one that required a different set of solutions to combat. For the first time in history, large numbers of people began dying in machine-related accidents. Other diseases had been amplified by human cultural innovation: dense cities with poorly designed waste removal allowed cholera to thrive, as we saw in chapter 3. But the mechanical carnage of the industrial age followed a different pattern. We invented a series of technologies designed for a specific purpose—steam-powered looms, rail locomotives, airplanes, automobiles—that turned out to have an unintended consequence: these inventions had the irritating habit of killing the people who used them.

Who was the first person killed by a machine? On this the historical record is by definition blurry. Do we consider a rifle a machine? A cannon? A catapult? The first person killed by a machine not explicitly designed for war was probably an employee in the Lancashire mills during the early days of the Industrial Revolution. It must have been shocking to see at first. Machine-based accidents introduced a kind of spectacular violence that had previously been witnessed only on the battlefield. Skulls were crushed, limbs severed; explosions turned bodies into unrecognizable biomass.

Before the automobile brought such carnage to everyday life, the railroad was the most visible source of machine-based accidents. Some of the first photographs to run in newspapers displayed the gruesome scenes of rail tragedies, the death toll hyped in oversized type. Charles Dickens barely escaped death in a rail accident in 1864, as he was nearing the end of writing his last masterpiece, *Our Mutual Friend*. (After extricating himself from the coach, he realized he had left the manuscript on the train and clambered back in to retrieve it.) The incident was said to have scarred him for the rest of his life.

The passengers were the lucky ones. Few jobs in the history of human employment have been more life threatening than being a railway worker in the middle of the nineteenth century. Just under 10 percent of workers in the so-called running trades—particularly those involved in coupling and decoupling the cars—experienced a serious injury each year. Anyone involved in the business could see it was killing people at an alarming rate. Rail titans like George Westinghouse introduced safety measures like passenger train air brakes, while Eli Janney invented a method for coupling trains automatically. But as is so often the case, it took statistics to shine sufficient light on the problem for outsiders to take notice. In 1888, the nascent Interstate Commerce Commission began gathering data on railroad accidents in the United States. The numbers they eventually released were scandalous: railway workers had a 1 in 117 chance of dying in an industrial accident.[2]

The data led directly to the passage of one of the most under-appreciated pieces of legislation in the history of the United States: the Safety Appliance Act, which compelled railroad companies by law to install power brakes and automatic couplers on all their trains. Within a decade, the efficacy of the state's intervention was undeniable: the mortality rates of rail workers were cut in half.

The Safety Appliance Act might sound to the modern ear as

though it had been designed to protect us from our washing machines, but it was a landmark nonetheless: the first American law passed with a primary focus on improving safety in the workplace. Hundreds of laws dedicated to reducing the threats posed by machines would follow in its wake.

Most of them would be targeted at automobiles.

EXACTLY HOW MANY human lives were sacrificed to the twentieth century's love affair with the automobile? Global numbers are difficult to estimate, but in the United States, accurate records have been kept since 1913. In little more than a century of driving, more than four million people have died in car accidents. Three times as many Americans have died in automobiles than died in all military conflicts going back to the Revolutionary War. (This figure surely undercounts the mortality effect of the automobile, as it does not include the environmental effect of air pollution and lead poisoning that were also side effects of a car-centric culture.)

Does any invention of the twentieth century—even those designed for combat—have a body count that rivals that of the automobile? The atomic bomb killed a hundred thousand; all airplane crashes combined add up to roughly the same number. At the peak of Hitler's Final Solution, Zyklon B and the gas chambers killed far more people than automobiles did over the same period. But measured over the course of the century, only the machine gun rivals the automobile as a mass killer.

The impact of automobile deaths on life expectancy was particularly intense because so many of the deaths involved young people. One way to register how extensive that death toll was is to note how many celebrities died before the age of fifty in car accidents:

the musicians Harry Chapin, Marc Bolan, and Eddie Cochran; the dancer Isadora Duncan; the writers Margaret Mitchell, Albert Camus, and Nathanael West. Members of royal families died tragically young in widely covered accidents, most famously Queen Astrid of Belgium and Princess Diana. The fathers of Bill Clinton and Barack Obama both died in automobile accidents at an early age. Car crashes took the lives of the actors Jane Mansfield and Paul Walker. But few automobile deaths resonated with the general public as widely as the 1955 death of James Dean, at the age of twenty-four, who was killed when his Porsche Spyder collided with a Ford Tudor at an intersection in central California.

At the time of Dean's death, almost every car manufactured offered only minimal safety features. Seat belts were practically non-existent and rarely worn; recessed steering wheels and crumple zones were unheard of; air bags and antilock-brake systems hadn't been invented yet. The Chevrolet Bel Air, 1955's bestselling family car, had no headrests, no rearview mirrors, no padding on the dashboard, and no seat belts. And yet the Interstate Highway Act and the postwar boom meant that millions of Americans were now regularly traveling at freeway speeds in cars that were astonishingly deadly in the event of a collision. With few exceptions, the automobile industry responded to the growing body count by throwing up its hands. Automobile fatalities were inevitable, they argued. It was simple physics. The forces in a crash were too great, and the human body was too fragile.

External innovations—traffic lights, speed limits—had lowered the odds of dying in a crash compared to the early days of driving. In 1935, there were fifteen fatalities for every 100,000 miles driven in the United States. By the time James Dean died in that Porsche Spyder, the fatality rate was half that. But the idea of reducing that number further by altering the design of the car itself was simply not part of the conversation. It wasn't that the automobile manufacturers were struggling to invent new safety innovations and just hadn't figured them out yet. The limitation was conceptual not technical. They were convinced that traveling at fifty miles per hour in a metal container was just fundamentally dangerous. (In this, the carmakers were not so different from the pessimists of the nineteenth century who surveyed the death toll of the new industrial hubs and concluded that cities on that scale, with that density, were fundamentally unhealthy.) The first breakthrough required to get out of that impasse was not a mechanical invention, but a way of seeing around the blind spot of the age. What was needed was not a solution to the problem, but a more fundamental shift: the belief that the problem could be solved in the first place.

Perhaps the most important early figure to embrace that belief was a Brooklyn-born pilot and engineer who managed to gain a revolutionary perspective on the problem of car safety through an experience that almost took his own life: dropping out of the sky in an airplane.

ONE DAY IN 1917, Hugh DeHaven, then a twenty-two-year-old student pilot, took off on an aerial gunnery training session in Texas, overseen by the Royal Flying Corps, where DeHaven was a cadet. Something in the session went horribly awry, and DeHaven's plane collided with another aircraft participating in the training. DeHaven suffered severe internal injuries; everyone else involved in the crash perished. In the months of recovery that followed, DeHaven found himself dwelling on the different outcomes of the crash.[3] Why had he been spared? A more spiritually inclined survivor might have assumed there was some kind of divine intervention at work. But DeHaven had a secular explanation: something in the design of the plane had protected him.

His military career curtailed by the accident, DeHaven took to inventing as his main trade. (He patented a device for bulk packaging newspapers that made him a wealthy man in his mid-thirties.) But all the while, something about that accident in Texas lingered at the back of his mind. He had grasped a fundamental truth about all vehicle fatalities—whether planes, trains, or automobiles— that the way in which a vehicle's structure frames and protects its occupants has a dramatic effect on mortality rates in high-speed collisions. DeHaven called it packaging. Build the cockpit of a plane, or the chassis of a car, one way, and its occupants perish in a crash; but protect the package another way, and they survive.

In 1933, DeHaven experienced the second machine-based accident that would shape his career: a gruesome car crash in which a dashboard knob punctured a driver's skull. Whatever post-traumatic stress he might have experienced from this second encounter with machine violence he channeled into testing and refining his packaging idea.

DeHaven started with eggs. He transformed his kitchen into a crash-impact laboratory, with layers of foam rubber lining the floor. He would drop eggs from ten feet on varying levels of foam, recording which materials kept the eggs from breaking on impact. Eventually the ceiling height of his kitchen began restricting his work; he then started dropping eggs from buildings in experimental packages designed to reduce the force of the impact on the ground. (Many high school physics classes today feature egg-drop competitions based on DeHaven's original research.) By the 1940s, he could drop an egg from the top of a ten-story building without damaging the shell.

Alongside the egg-drop experiments, DeHaven collected news reports of car accidents, with an extra focus on the cases where someone had survived a high-speed crash. He also curated stories of suicide attempts and other accidental falls where people had miraculously survived a free fall of more than a hundred feet. He calculated the physics of these collisions, ultimately determining that the human body was capable of surviving g-forces that were two hundred times more powerful than ordinary gravity on earth. If you could keep passengers from being impaled on the steering wheel, or flying through the windshield, high-speed accidents need not be death sentences. DeHaven collected this research into a paper, published in 1942, called "Mechanical Analysis of Survival in Falls from Heights of Fifty to One Hundred and Fifty Feet." The paper focused primarily on eight case studies of improbable free-fall survivors, noting the circumstances, injuries, and g-forces at work in each event:

> A woman, who jumped from a 17th floor, falling 144 feet (43 meters) in a similar "steamer chair" position, landed on a metal ventilator box 24 inches (61 cm) wide, 18 inches (46 cm) high and 10 feet (300 cm) long. The force of her fall crushed the structure to the depth of 12 to 18 inches (30 to 46 cm). Both arms and one leg extended beyond the area of the ventilator, with resultant fractures of both bones of both forearms, the left humerus and extensive injuries to the left foot. She remembers falling and landing. There were no marks on her head or loss of consciousness. She sat up and asked to be taken back to her room. No evidence of

abdominal or intrathoracic injury could be determined, and roentgen examination failed to reveal other fractures. The average gravity increase was a minimum of 80 g and an average of 100 g.[4]

DeHaven's paper made for unusual reading. Tales of miraculous survival that would normally be printed in 100-point type on a tabloid cover—WOMAN SURVIVES FALL FROM 17TH FLOOR!—were recounted in clinical detail. And while it might have seemed, on the surface, to offer advice to would-be jumpers, DeHaven made his ultimate objective clear in the final lines:

> The human body can tolerate and expend a force of two hundred times the force of gravity for brief intervals during which the force acts in transverse relation to the long axis of the body. It is reasonable to assume that structural provisions to reduce impact and distribute pressure can enhance survival and modify injury within wide limits in aircraft and automobile accidents.[5]

Translated into a language that ordinary car owners could understand, DeHaven's words were revolutionary: the occupants of a car colliding with another vehicle at fifty miles per hour were not doomed by physics to die in the crash. The right packaging could have them walking away from the accident unscathed. DeHaven's paper marked the origin of a new field: injury science. In the words of a later practitioner in that field, the paper introduced the radical idea that "crashes and their resultant injuries were not inevitable but rather were predictable and, therefore, preventable events."

DeHaven had made his argument with eggs and algebra and

newspaper clippings. But sometimes a different kind of persuasion is required to change the conventional wisdom. In the story of auto safety, that mode of persuasion is best exemplified by Colonel John Stapp. Stapp was a classic polymath: a surgeon, biophysicist, and pilot. He was known for a time as the "fastest man on earth." The nickname was somewhat ironic, given that Stapp's lasting contribution to automobile and airplane safety lay in understanding the physics of radical decelerations. He made headlines as a speed demon, but his real legacy was all about what happens to the human body when it slows down.

ON NOVEMBER 14, 1947, John Stapp, then a thirty-seven-year-old project officer at the US Army Aeromedical Research Laboratory, rose to the podium in the ballroom of Boston's Statler Hotel to deliver an address to an annual convention of military surgeons. His talk—later published as a short paper called *Problems of Human Engineering in Regard to Sudden Decelerative Forces on Man*—belonged to a crucial genre in the history of science, one that is often neglected: a work that does not suggest a new answer or explanation but rather identifies a new kind of problem worthy of exploration. The problem, put simply, was trying to figure out what happened to the human body when it went from going one hundred miles per hour to zero in a few seconds or less. This was, Stapp noted, a genuinely new sort of problem, one that had been made possible by recent technological developments. And while he was addressing an audience of physicians, Stapp argued that it was a problem that could be productively approached through the lens of engineering. "Until the demands of modern aviation began to exceed the limits of human tolerance for acceleration and deceleration,"

Stapp explained, "medical men had too little knowledge of or interest in engineering to apply to the problem of the physiological and structural stress analysis of the human body." Early in the speech, Stapp outlined the challenges of the approach:

> To the human engineer, man is a thin, flexible leather sack filled with thirteen gallons of fibrous and gelatinous material, inadequately supported by an articulated bony framework. Surmounting this sack is a bone box filled with a gelatinous matter which is attached to the sack by means of a flexible coupling of bony and fibrous composition. The center of gravity of this irregular mass depends on the position of four articulated appendages of bony and fibrous structures. Fuel and lubricants are conveyed to all parts of this machine by flexible hydraulic systems with low pressure tolerances actuated by a central pump. Because of its irregular shape, variety of materials, and composition . . . the stress analysis of this machine for externally applied forces is exceedingly complex.[6]

To combat that complexity, it was necessary to go beyond the egg drops and case studies that Hugh DeHaven had relied on in his groundbreaking paper five years before. "The problem is not simple," Stapp noted with a twinkle in his eye. "We cannot tie a microphone to a subject, drop him out of successive floors of a building down the elevator shaft, and assume his outcries to be proportional to the effect of the forces." Instead, Stapp explained, the Aeromedical Lab had developed a new kind of technology that would perform stress analysis on actual human bodies—as well as "anthropomorphic dummies"—experiencing the radical deceleration of an airplane crash. They called it the linear decelerator, but most iterations of the machine that would follow went by another, more memorable name: the rocket sled.

It was an apt nomenclature. The machines were effectively a pack of solid rocket-fuel motors strung on the back of a sled carrying a single passenger, usually sitting upright and strapped into a padded chair. The whole contraption slid over precisely aligned rails to keep it from veering off in random directions. (There were no wheels involved.) The brake systems were robust enough to bring a sled traveling at 120 miles per hour to a standstill in just a few seconds. Early versions—like the linear decelerator that Stapp first built at the Aeromedical Lab—could hit top speeds in the 200s.

Stapp was not just a designer; he was an active user of the device. Over the years he broke ribs, fractured his wrist twice, suffered temporary vision loss. But each time he rode the device, a small battalion of sensors were dutifully taking notes on the slightest changes in his body as it battled those prodigious g-forces. That was the nature of a stress analysis: someone had to get stressed in the process if you couldn't build accurate enough crash dummies. This is what makes Stapp such a fascinating figure: he was providing both the stress *and* the analysis.

John Stapp is now mostly remembered for his involvement with a contraption that debuted in 1954: the rocket sled called Sonic Wind 1. On December 10, 1954, Stapp made history on the Holloman High Speed Test Track in New Mexico, by riding the Sonic Wind at a peak speed of 628 miles per hour, before slamming to a death-defying halt in just 1.4 seconds.[7] Stapp's courage here is not to be undervalued. It was not at all clear that traveling over land at near the speed of sound was a survivable experience. While he was strapped into a throne-like seat with carefully positioned restraints,

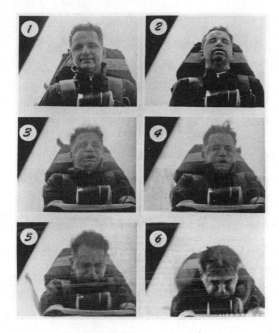

John Stapp's deceleration
test

he performed the test without any kind of protection over his face.
The still frames capture some of the physical forces that were at
work on John Stapp in those two seconds of deceleration. Look at
the difference in his face between frame 1 and frame 6. He appears
to have gained fifty pounds in a matter of seconds: all the "gelati-
nous matter" surging forward as his spine and torso retract at prodigious speeds. He looks twenty years older in that final frame, as
though the physics of aging were somehow a tape that you could
play at a much higher speed.

No human being had ever traveled on land at anywhere near that
speed. Stapp was instantly declared the fastest man on earth, appearing on the cover of *Life* magazine—the pinnacle of American
media during that period—shortly after his legendary ride on
Sonic Wind 1. The test had been officially designed for aeronautical
purposes; the air force wanted to know if it even made sense to

offer ejection seats on supersonic aircraft, given the wind speeds pilots would experience in the ejection process. The answer to that question is visible in those images. It wasn't pretty—Stapp again lost vision temporarily, and his face was badly bruised—but he was alive at the end of it and suffered no permanent injuries. "I felt a sensation in the eyes," he later recalled, "somewhat like the extraction of a molar without anesthetic."[8] But he had survived.

That was good news for the small number of people who would fly at supersonic speeds in the next decade. But it was also good news for the millions who used a more conventional form of transportation. If you could decelerate from 600 miles per hour to zero in a matter of seconds without major injuries, surely you should be able to survive a collision at sixty miles per hour. In his years in the air force, Stapp had noticed that more of his fellow servicemen died in automobiles than in airplanes. And so in May 1955, Stapp invited twenty-six people involved in the automobile industry to visit Holloman Air Force Base to witness the rocket sled in action and discuss ways in which the lessons of Stapp's research could be applied to auto safety. The sessions were repeated the next year; more than sixty years later, the Stapp Car Crash Conference is still the main industry meeting for the extended community of auto safety experts.

Stapp also directly advised Ford on the design of the 1956 Fairlane Crown Victoria, which featured the special Lifeguard safety package. (The safety features were a passion project for then Ford executive Robert McNamara, one of the only auto executives at the time to show any interest in reducing fatalities.) For the first time, an automaker was attempting to compete based on the safety record of its vehicles, and not just their styling or horsepower. The Crown Victoria featured safety door latches, a lap belt, padded dashboard, padded sun visor, and a recessed steering wheel. But Ford's more powerful competitor General Motors felt that highlighting the dangers of driving could be catastrophic for the entire industry. They threatened to take Ford to court, and for whatever reason, the Lifeguard package wasn't a hit with consumers. Watching the dismal sales figures come in, Henry Ford II complained to a reporter, "McNamara is selling safety, but Chevrolet is selling cars."[9] Hugh DeHaven and John Stapp had convincingly undermined the consensus that simple physics limited our ability to make automobile crashes safer. But a new consensus quickly emerged to replace it: safety didn't sell.

DESPITE THE TENACIOUS EFFORTS of DeHaven and Stapp, the first meaningful breakthrough in auto safety—to this day, the one with the largest impact—would come not from Detroit but from Sweden. In the mid-fifties, Volvo hired an aeronautical engineer named Nils Bohlin, who had been working on emergency ejection seats at Saab's aerospace division. Bohlin began tinkering with a piece of equipment that had been largely an oversight in most automobiles up until that point: the seat belt. Many cars were sold without any seat belts at all; the models that did include them offered poorly designed lap belts that offered minimal protection in the event of a crash. They were rarely worn, even by children.

Borrowing from the approach to safety restraint used by military pilots, Bohlin quickly developed what he called a three-point design. The belt had to absorb g-forces on both the chest and the pelvis, minimizing soft tissue stress under impact, but at the same time it had to be simple to snap on, easy enough that a child could master it. Bohlin's design brought together a shoulder and lap belt that buckled together in a V formation at the passenger's side, which meant the buckle itself wouldn't cause injuries in a collision. It was an elegant design, the basis for the seat belts that now come standard on every car manufactured anywhere in the world. An early prototype of the shoulder strap had decapitated a few crash dummies, which led to a rumor that the seat belt itself could kill you in a crash. To combat those rumors, Volvo actually hired a race-car driver to perform death-defying stunts—deliberately rolling his car at high speeds—all the time wearing Nils Bohlin's three-point seat belt to stay safe.

By 1959, Volvo was selling cars with the three-point seat belt as a standard feature. Early data suggested that this one addition was single-handedly reducing auto fatalities by 75 percent. Three years later, Bohlin was granted patent number US3043625A by the US Patent and Trademark Office for a "Three-point seat belt systems comprising two side lower and one side upper anchoring devices." Recognizing the wider humanitarian benefits of the technology, Volvo chose not to enforce the patent—making Bohlin's design freely available to all car manufacturers worldwide. The ultimate effect of Bohlin's design was staggering. More than one million lives—many of them young ones—have been saved by the three-point seat belt. A few decades after it was awarded, the Bohlin patent was recognized as one of the eight patents to have had "the greatest significance for humanity"[10] over the preceding century.

Even with a clear track record of decreased fatalities and an open patent, the Big Three American car companies continued to resist prioritizing safety in their vehicle design through the first half of the 1960s. In the end, they were compelled to change their ways not by egg-drop experiments or rocket sleds, but rather by the journalist and lawyer Ralph Nader. Until playing the spoiler role in the 2000 Presidential election, Nader was most recognized for his 1965 bestseller, *Unsafe at Any Speed: The Designed-In Dangers of the American Automobile.* The opening line of the book offered a sobering assessment of the car's effect on society: "For over half a century the automobile has brought death, injury, and the most inestimable sorrow and deprivation to millions of people."[11] In the book, Nader praised the visionary experiments of DeHaven and Stapp, and excoriated the auto companies for ignoring what he called "a gap between existing design and attainable safety." In the opening chapter, he set his sights on GM's Chevrolet Corvair, which he memorably derided for its propensity for "one-car accidents." (A poorly engineered suspension system made it easy for the driver to lose control of the car and, on numerous occasions, flip over—even without any contact with another vehicle.)

Even before the book was published, GM had hired a private investigator to dig up dirt on Nader. He received odd phone calls in the night; women tried to seduce him at coffee counters; friends and colleagues were questioned on the pretense that Nader was being considered for a new job, and asked questions about his sex life and involvement with left-wing political groups. Eventually GM president James Roche was brought before a Senate committee and forced to apologize publicly for its campaign of harassment against the young activist, further propelling the sales of Nader's book.

The impact on popular opinion—on Main Street and inside the Beltway—mirrored the sudden shift that followed the thalidomide crisis a few years before. Senator Abraham Ribicoff, who led the hearings into the GM harassment campaign, declared that traffic accidents were a "new type of social problem that springs from affluence and abundance rather than from crisis and convulsion."[12] In September 1966, with the support of President Lyndon Johnson, Congress enacted the National Traffic and Motor Vehicle Safety Act, with the aim of providing "a coordinated national safety program and establishment of safety standards for motor vehicles in interstate commerce to reduce traffic accidents and the deaths, injuries, and property damage which occur in such accidents." The act radically expanded the government's regulatory oversight over the auto industry and had wide-ranging and complex implications. It would eventually lead to the formation of the US Department of Transportation. But the most important one was easy enough to understand: for the first time, every new car sold in the United States had to come with seat belts installed. Just a decade before, seat belts had been dismissed as a folly, an inconvenience—or worse, a potential threat in their own right. Now they were the law.

SHORTLY AFTER THE 1966 ACT was passed, the Speaker of the House, John McCormack, credited the legislation's success to the "crusading spirit of one individual who believed he could do something . . . Ralph Nader."[13] In a way, Nader was following a playbook that extended back to the early muckrakers—to Jacob Riis and Upton Sinclair, even to Charles Dickens—using the power of journalism to change the general public's attitude toward a crucial social problem, and compel lawmakers to enact legislation to address the problem. Nader's true innovation was to shift the focus from workers to consumers. Sinclair and his ilk had targeted the work environments of factories and slaughterhouses and other sites of industrial-age labor. If they had arguments with Detroit, they revolved around the assembly line workers: their wages, hours on the job, occupational hazards. *Unsafe at Any Speed*, on the other hand, was a book aimed at protecting the people who bought the cars, not the people who made them. Nader's key contribution lay in inventing a whole new kind of political figure, a Frank Leslie for the television age: the consumer advocate, using the media and the courts to compel the private sector to make safer products.

But as important as Nader was to the 1966 act, the movement to "buckle up" involved a much wider range of participants. As usual, the main players came from a diverse range of backgrounds: a maverick inventor, a daredevil pilot, an aviation engineer, a firebrand lawyer, and the United States Congress. They used a mix of tools to make the case that automobile safety could be improved: egg drops and rocket sleds, stunt drivers and bestselling books. In this they followed the pattern that we have seen over and over again in the preceding chapters. Real change often requires a first step of convincing people that the existing problem is not inevitable; and devising a solution requires a diverse network of talents, building on one another's work.

The most striking thing about the story of car safety, though, is the one group that is almost entirely missing from the list of the seat belt's main proponents: the automotive industry itself. With the exception of Nils Bohlin and Volvo, none of the key events that made buckling up second nature to us today came from the automobile establishment. Progress did not come about "naturally" by allowing the private sector to innovate, making safer products because they would logically appeal to consumers. Instead, the progress had to be fought for by outsiders, battling opposing forces to make the case for it. Some of those opposing forces were a matter of physics; some took the form of private investigators hired by General Motors.

The seat belt, of course, was just one of a series of safety innovations that are now standard components of the automotive environment today. In the decades that followed *Unsafe at Any Speed*, the car companies did become more committed to advancing safety innovation themselves, though progress continued to be driven by outsiders as well. The airbag, originally invented in the 1950s, was refined by a number of engineers until becoming mandatory in 1989. Antilock brakes, pioneered by the airline industry, became standard in cars in the 1990s. Activists working in the mode of Ralph Nader continued to drive change. The tragic death of her daughter in a drunk driving accident compelled Candace Lightner to form Mothers Against Drunk Driving (MADD) in 1980, leading to a radical decrease in alcohol-related accidents. Celebrity deaths also played a role. After Princess Diana died while not wearing a seat belt in the rear seat of the Mercedes she was traveling in, backseat seat belt use rose by 500 percent in the UK, and more than doubled in the United States.

What was the total impact of all these inventions and interventions? If you sit behind the wheel of an automobile today, you are more than *ten times* less likely to die than you would have been when automobiles first became part of modern life. Recall that car accidents were the third common cause of death when James Dean stepped into that Porsche Spyder. Today they are not even in the top ten.

Consider the chart on the next page that shows the decline in US fatalities per 100,000 miles driven from 1955 to the present day.[14]

The most pronounced drop in mortality comes in the five years after the passage of the 1966 legislation, as seat belt use becomes

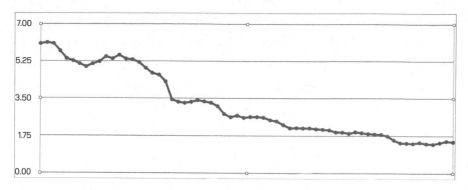

US automobile fatalities per 100,000 miles driven, 1955–2018

increasingly common and top speed limits are reduced nationally to fifty-five miles per hour. But the most striking thing about the chart is the steady, incremental improvements in safety that occurred in the following three decades. There are no sudden dramatic improvements; each year, with a few exceptions, is just slightly safer than the year before. That's the kind of chart you see when progress comes not from one genius inventor or dramatic breakthrough but the work of thousands, each attacking the problem from different angles: consumer advocates, industry engineers, government regulators, grief-stricken mothers. Because each year is just a fraction better than the one before, we never hear about the improvement. Celebrity deaths and other tragic accidents continue to make headlines, but the lives saved never make it to the front page because year by year the changes are small. But when you stack them up over a century of driving, they're miraculous.

All those innovations and legal reforms were ultimately about one thing: How can we keep people safer in the event of a crash? The technology has changed, but the nature of the problem is the same one that Hugh DeHaven first began wrestling with in 1917 after his near-fatal accident as a young cadet. But in recent years, a new possibility has emerged, one as radical as DeHaven's argument about survivability was back in the 1940s. Could we design cars capable of avoiding accidents altogether? This is the dream of the self-driving car, powered by machine-learning algorithms and elaborate sensors that help the car make assessments of complex, changing road conditions far faster than humans are capable of. The radical increase in safety introduced by the seat belt was a matter of understanding the physics of a crash. The self-driving revolution, its proponents believe, will revolve around data. With smart-enough cars—perhaps with digital coordination between those cars—accidents themselves could potentially become as rare as plane crashes have become in recent years. Perhaps not surprisingly, the main actors driving this new paradigm are not based in Detroit but are instead in Silicon Valley, at companies such as Google and Tesla.

This potential safety revolution requires extensive training sets, given the variability of real-world driving conditions. It also requires algorithmic decision-making. The "packaging" that protects you is no longer just the airbags or the collapsing steering wheel but rather the car's ability to make the right choice at the right time. Already cars manufactured by Tesla monitor every mile driven with humans at the wheel, recording and learning from everything that happens. The car analyzes the choices the driver makes while traveling—swerving to avoid a pedestrian, tapping the brakes to signal the car behind to back off, slowing down in foggy weather. All the while the car quietly makes its own simulated decisions, comparing its imaginary driving to the driver's actual performance. Through this kind of study, over time, advocates believe machine learning will train our cars to become far better drivers than we are ourselves.

Even if this scenario does come to pass, handing over our driving decisions to an algorithm will create strange moral dilemmas. What happens if the car confronts a situation where it has to choose between potentially risking its driver's life versus running over two pedestrians? Which life should it be programmed to value? If the self-driving revolution happens, we may well see the number of deaths per 100,000 miles drop to near zero. But in the process something bizarre will have happened: cars will have to possess something like ethics. Some will have more aggressive settings; they'll be more open to risk. Others might be programmed to prioritize pedestrian safety over other drivers. Perhaps this is a natural evolution. We used to choose a car based on the design of its fins or its zero-to-sixty acceleration. But in the future, some of us may choose to buy a car based on its moral values.

In that self-driving future, there will undoubtedly be edge cases where an automobile has to make one of those impossible decisions, choosing between killing one person or another. Those incidents will doubtless attract headlines and trigger outrage, even if the sum total of human fatalities has been greatly reduced by our algorithmic drivers. Those incidents will also mark a milestone in the centuries long history of people being killed by the machines they invent. Ever since Mary Ward perished in Ireland, crushed by that steam-powered road locomotive, such deaths have been classified as accidents. But what category do you use when a machine kills a human because it decided to?

8

FEED THE WORLD

THE DECLINE OF FAMINE

Roughly three decades ago, the biologist and complexity theorist Stuart Kauffman coined a phrase to describe the way meaningful change happens, in both natural and cultural systems. Each new change—the evolution of bipedal walking, for instance, or the invention of the printing press—unlocks new doors of possibility for other changes, Kauffman noted. Our ancestors start walking on two feet, which frees up their hands for other kinds of activity, which leads to the evolution of opposable thumbs. The printing press creates the possibility of storing and sharing scientific insights, which leads to the invention of new citation systems like page numbers and footnotes, which eventually leads to the idea of hyperlinks many centuries later. Kauffman gave those secondary effects a memorable name: the "adjacent possible."[1] New scientific

breakthroughs changed the world not just through the new func-
tionalities they introduced, but also in the ways they expanded
the adjacent possible: through the lateral effects they created, the
new ideas that suddenly became thinkable because of them. An
FDA that demanded proof of efficacy for new drugs was not part of
the adjacent possible in 1937, during the Elixir Sulfanilamide crisis,
because RCTs had not been invented yet. But by the time Frances
Oldham Kelsey began looking into thalidomide, a more rigorous
standard was, in fact, available to her and her colleagues, thanks to
the work of Austin Bradford Hill and Richard Doll. The invention
of the RCT created a new template for experiment design, but it
made a new kind of regulatory intervention possible as well. Simi-
larly, antibiotics opened the door for new elective surgeries by
greatly reducing the risks of fatal infections.

The strange thing about the adjacent possible is that the new
doors unlocked by each innovation do not always seem, at first
glance, to be all that adjacent. Big changes in society often happen
because a new idea in one field triggers changes in a seemingly unrelated field. Intellectual histories, for understandable reasons, tend
to underemphasize these kinds of causal leaps; the history of chemistry focuses on the chemists, while the history of epidemiology
focuses on the epidemiologists. But the truth is that the new ideas
introduced in these fields have a tendency to leap over these disci-
plinary barriers. Gutenberg's printing press borrowed a key piece of
technology from winemakers, who had developed what was called
a screw press for crushing grapes. The winemakers had no idea they
were opening space in the adjacent possible for a publishing revolution, but that is exactly what their technology ended up doing.

Because the story of human life expectancy is tied to so many
different kinds of innovations—statistics, chemistry, new modes of

government oversight—it should come as no surprise that the story features many unlikely links of causation, the health equivalent of the winemakers unwittingly helping to jumpstart the Gutenberg age. Consider this question: What new idea or technology discovered in the nineteenth century had the biggest impact on life expectancy in the twentieth? A few obvious candidates come to mind, some of which we have already explored: Farr's surveillance and statistics revolution; the concept of waterborne diseases. But you can also make an argument that the single most influential idea came from a much more surprising place: the discovery that soil is alive.

How we got to that understanding was complicated, but it largely happened in a burst of cross-disciplinary activity in the middle decades of the nineteenth century. Scientists began to realize that soil wasn't just a bunch of ground up rocks, inert and unchanging. It had a metabolism. It required energy inputs and waste management. In the right circumstances it could be capable of staggering fecundity; in the wrong, it could wither into lifeless dust. And it was teeming with microscopic life-forms, each playing a critical role in what we now call the nitrogen cycle.

The most important stage in this cycle was the "fixing" of nitrogen, converting it from atmospheric nitrogen into ammonia nitrates that plants use as food. The problem with nitrogen is that, while the gas is abundant in our atmosphere, it doesn't combine easily with other elements in its standard state. Over billions of years of evolution, soil ecosystems had overcome that limitation through the dedicated work of microorganisms known as diazotrophs, which convert nitrogen into ammonia that can then be used to fuel plant growth. Other microorganisms specialize in decomposing plant and animal life, releasing ammonia as well. Seen from this new perspective, soil suddenly started to seem like a chemical production plant, with untold millions of microscopic workers laboring away to produce nitrates, literally pulling them out of thin air. That knowledge proved to be essential for the team of microbiologists, chemists, and agronomists desperately trying to devise a way to mass-produce antibiotics in time to win the war.

The antibiotics revolution required several critical breakthroughs to graduate from a promising but enigmatic mold to a global magic bullet. One of them was surely the development of modern soil science. The scientists exploring the invisible kingdom beneath our feet had no idea they were establishing the building blocks of the twentieth century's most critical medical innovation. If you'd asked them, they would have said their research had nothing to do with medicine at all. But that is the way the adjacent possible works: sometimes the new doors that are unlocked take you to unexpected places. Sometimes they drop you in a completely different wing of the building.

The discovery that simple dirt had a complicated metabolism would have another, slightly more predictable effect on life expectancy. Understanding that soil was alive helped us ward off infections, but it also helped us ward off another omnipresent threat: hunger.

ON MAY 20, 1915, as the first battles of World War I raged across Europe and the Middle East, an American diplomat named Ralph G. Bader, stationed in Tehran, wrote a dispatch back to the states reporting on how the then-neutral nation of Persia was responding to the global disruption of the war. Goods imported from Europe had increased dramatically in value, Bader reported, but the indigenous food supply had largely been unaffected. "The cost of living for the native population, whose principal articles of diet are mutton, rice and bread made from whole wheat flour, has only slightly increased," he wrote. But by October, with the Russians, Turks, and British battling for control of the country, ominous signs had begun to appear. Jefferson Caffery, the American chargé d'affaires, reported that breadlines had become ubiquitous on the streets of Tehran, and the price of sugar had increased from ten cents a pound to more than a dollar a pound in a matter of months.

The disturbance of normal food distribution networks caused by foreign invaders was then exacerbated over the next year by a severe drought that settled in across wide swaths of Persia. A Kansas-born lawyer named John Lawrence Caldwell was the US ambassador to Persia during that period. By 1917 he was reporting that riots had commenced. "It cannot be doubted that deaths and starvation will multiply this winter," he warned. A later telegram from Caldwell documented the stratospheric rise of prices: staples such as rice had increased from five cents a pound to as much as two dollars a pound. It was the equivalent of walking into a present-day supermarket and discovering that a gallon of milk now costs $200. (Interestingly, Caldwell noted that the primary problem appeared to involve the cost of goods and not their availability. "Wheat costs from fifteen to twenty dollars per bushel," he observed, "though quite a supply could be had at these prices.")[2] With the core dietary needs of the Persian population effectively unaffordable, a devastating famine began to sweep across the nation. A professor at the American College telegrammed back to the states in 1918: "40,000 destitute in Tehran alone. People eating dead animals. Women abandoning their infants."

When the British major general L. C. Dunsterville arrived in Persia around this time, commencing a British military occupation that would last for three years, he encountered a country on the verge of utter collapse. His subsequent recollections of the experience shift effortlessly between his genuine horror at the human suffering and his grotesque stereotypes about the "typical" Oriental:

> The evidence of famine was terrible and in a walk through the town one was confronted with the most awful sights. Nobody could endure such scenes if he were not endowed with the wonderful apathy of the Oriental: "It is the will of God!" So the people die and no one makes any effort to help, and a dead body in the road lies unnoticed until an effort to secure some sort of burial becomes unavoidable. I passed in a main thoroughfare the body of a boy about nine years of age who had evidently died during the day; he lay with his face buried in the mud, and the people passed by on either side as if he were merely any ordinary obstruction in the roadway.[3]

As if the casual racism of Dunsterville's notion of the "wonderful apathy of the Oriental" weren't enough, evidence now suggests that a major cause of the price increases that triggered the famine came from the British army's purchase of massive food stocks to feed its troops throughout the Middle East. To Dunsterville's imperial mindset, the "awful sights" of famine in the streets and countryside of Persia appeared as so much evidence of a country incapable of managing its own affairs. In reality, Dunsterville was there to "rescue" the Persians from a crisis the British themselves had helped precipitate.

The ultimate cost in human lives from the Great Persian Famine of 1916–18 remains a matter of debate. The population of Tehran appears to have been cut in half during the peak years of the famine, dropping from four hundred thousand to two hundred thousand. Some historians have argued that mortality rates nationwide were comparably dramatic, as high as 50 percent. Others believe that closer to 20 percent of the population died of hunger during those three turbulent years.

As devastating as the Great Persian Famine was to the people of the region, it was only the beginning of a wave of catastrophic famines that would race around the world over the ensuing decade. The death toll from starvation during this period almost certainly exceeded the lives lost to military conflict in the war. Only influenza managed to do more damage. Upward of fifty million people died during the famines of the 1920s—some triggered by unusual weather patterns, some by disturbances in food distribution caused by the war, some by the disastrous initial experiments with central planning that had begun in the newly formed Soviet Union.

Though these numbers seem staggering to us now, as a percentage of the overall population the mortality rate from famine during that decade was not unusual compared to food crises throughout human history. The famous Irish potato famine of the late 1840s killed roughly one eighth of the population and forced another quarter of the population to emigrate in search of food to other regions of the world, mostly the United States. The beginnings of what we now call the Little Ice Age in the 1300s brought floods and unusually cold weather to northern Europe, producing famines that may have taken the lives of as much as a third of the population. Scholars now believe that the mysterious collapse of the Mayan civilization was partially triggered by an extreme drought between AD 1020 and 1100 that led to massive crop failure, ultimately causing the advanced Mesoamerican culture to vanish practically overnight.

Hieroglyphs discovered on an island in the Nile River tell the story of a seven-year famine that brought chaos and political unrest to the reign of the Egyptian pharaoh Djoser, three thousand years before the birth of Christ.

Mass famines have been a nearly unavoidable corollary of agricultural societies; in the ultimate tally, they may well have taken more lives than warfare over the course of human history. In the modern era, where we have reasonably accurate assessments of the death tolls, famine seems to have been the more deadly force: more than 120 million people are believed to have died from famines worldwide between 1870 and 1970, likely a few million more than the death toll from military conflicts.

Chronic food shortages have other, more subtle, costs even when they do not trigger mass death. As Robert W. Fogel has observed, diets of most eighteenth- and early nineteenth-century Europeans were the equivalent of diets in Rwanda or India in the 1970s, countries where a significant portion of the population suffered from chronic malnutrition. Those limited diets put a ceiling on the amount of labor that could be performed by the population. "At the end of the eighteenth century British agriculture, even when supplemented by imports, was simply not productive enough to provide more than 80 percent of the potential labor force with enough calories to sustain regular manual labor,"[4] Fogel writes. Those missing calories also had a material effect on overall health. In his seminal work, *The Modern Rise of Population*, Thomas McKeown attributed many of the nineteenth-century gains in life expectancy to improvements in diet, arguing that it was agricultural improvements, not medical ones, that triggered the first upward march of the great escape. "In Europe," McKeown argued, "there was a large increase in food supplies between the end of the seventeenth century and the mid-nineteenth, in Britain sufficient to feed a population that had trebled in size, without significant imported food."[5]

McKeown didn't use this exact language, but in a real sense what he was describing was an energy revolution: energy, in this case, measured in calories consumed and not steam power. A population living on the edge of starvation—without sufficient energy intake to maintain basic metabolic functions—is a population that will be more vulnerable to opportunistic infections, even if it manages to steer clear of mass starvation. When we talk about the energy revolutions of the nineteenth century, the steam-powered factory system naturally comes to mind, but in McKeown's model, the primary driver was "the more effective application of traditional methods—increased land use, manuring, winter feeding, rotation of crops, etc.—rather than the technical and chemical measures associated with industrialization."[6] We started living longer, in other words, because we became better farmers, not better doctors.

Attempting to assess the caloric intake of people who lived more than two centuries ago presents a challenge to demographic historians, given that there was no William Farr recording the regular diets of ordinary people during the period. But one key measure of childhood nutrition levels is adult height. Societies with chronically malnourished children produce much shorter grown-ups than societies where the children are well fed. When we see rapid changes in adult height between generations, changes in early childhood diets are almost always the reason. (The average Japanese millennial, for instance, is almost a head taller than his or her grandparents, thanks to improvements in diet after World War II.) Building on McKeown's argument, Fogel presented evidence that average heights in England did increase by about five centimeters between 1750 and 1900, suggesting some meaningful improvement in diet during that period.

Agricultural improvements were not sufficient to allow Europe to escape from the ancient threat of mass starvation during that period. A famine in France helped trigger the French Revolution in the late 1780s; multiple famines in Scandinavia killed hundreds of thousands during the late 1800s; and, of course, the Irish potato famine resulted in casualties numbering more than a million. Globally, starvation would continue driving down human life expectancy significantly until the 1970s.

And then, almost overnight, famine released the ancient stranglehold that it has had on human society. Somewhere around five million people have died from famines between 1980 and today, compared to roughly fifty million over the preceding forty years. The drop is even more pronounced when you factor in global population growth during that period. Calculated on a per capita basis, famine deaths have declined from 82 per 100,000 people in the wake of the Persian Great Famine to just .5 per 100,000 people over the last five years.[7] Small-scale famines still happen, and there is every reason to believe that the deep-seated disturbances of climate change—both in terms of altered ecosystems and the demographic chaos of mass migration—will cause them to increase in the coming decades. But for the last forty years at least, the trend lines are about as encouraging as they can be. We have reduced the death toll of famine with something like the efficiency with which we reduced the death toll of tuberculosis: we transformed it from a looming threat, an unavoidable fact of life for many societies around the world, to a rarity, something that only 1 percent of the world's population need ever worry about.[8] It may be only a temporary peace, and the forces that drive mass starvation may come thundering back once the seas rise high enough. But it is a truce that has managed to last for forty years, with no sign of ending. Ironically, our peace with the great nemesis of famine was at least partially enabled by the technology of war.

FOR EONS, ammonium nitrate had been employed by plants all around the world as a natural fertilizer to support their growth. But a new use for the material opened up about a thousand years ago, when the Chinese first began experimenting with the explosive power of a close chemical relative, potassium nitrate, also known as saltpeter, the primary ingredient of gunpowder. Nitrogen itself was first isolated and named in the 1770s, during a period of rapid advances in chemistry. (Oxygen, too, was identified within a year of nitrogen's discovery.) By the nineteenth century, it had become clear that the nitrates could be used to encourage plant growth *and* blow things up. (Ammonium nitrate bombs are still used by terrorist groups, most notoriously in the 1995 Oklahoma City bombing.) But the ability to manufacture these nitrates had not yet become part of the adjacent possible. The only available option for humans who wished to utilize nitrates—at war or in their gardens—was to locate natural reserves of the chemical. And that is how seabird and bat excrement became one of the most highly prized commodities of the nineteenth century.

For more than a thousand years, the indigenous populations of coastal Peru made regular voyages to scrape what they called guano off the rocky terrain of nearby islands. The seabird waste turned what had been infertile desert into a thriving soil. The Inca empire shipped guano throughout South America to improve crop yields. In the early nineteenth century, Europeans finally recognized the commercial value of guano, after centuries spent paying more attention to the gold and silver of South America than its reserves of bat and seabird excrement. In 1840, Peruvians exploring the Chincha Islands off the country's southern Pacific coast made a discovery that rivaled the mythical El Dorado: deposits of guano caked into mounds more than 150 feet high, the single largest reserve of nitrates yet discovered. At this point, one could say that the world, quite literally, went batshit crazy. Whole regions were colonized; natural ecosystems disturbed; wars were fought. Farmers all around the world used Peruvian guano to increase the fertility of the soil. Bat guano from caves in the United States was a primary resource for gunpowder for the Confederate army during the American Civil War.[9]

The guano boom had an inevitable bust in its future because the bats and birds simply weren't producing enough waste to keep up with the demand. In the years preceding the outbreak of World War I, Germany became increasingly concerned about its ability to generate enough bombs to fight its European rivals. The limiting factor: dwindling supplies of nitrates, originally sourced from guano. The German chemist Fritz Haber began investigating ways to synthesize nitrates in the lab, but by 1908 he had perfected a system that could create ammonium nitrate without relying on diazotrophs or seabirds. It was alchemy of the first order: creating a valuable commodity out of simple air and heat (along with an iron catalyst). The chemist and industrialist Carl Bosch then designed a system where Haber's process could be reproduced at scale, with factories producing tons of ammonium nitrate. It is unclear how many deaths during the Great War might have been avoided had Haber and Bosch not teamed up to discover and amplify the technique of artificially "fixing" nitrogen. But it likely numbered in the hundreds of thousands, if not more.

Still, there was that strange property of nitrogen: it was as useful for the farmers as it was for the bomb makers. Once you could produce nitrates in a factory, the world of agriculture lost its dependence on naturally fertile soils or bat guano. Any field, however lifeless, could be supplemented by nitrates to jump-start the soil ecosystem. Figuring out how to make more bombs turned out to help us invent an entirely new concept: artificial fertilizer. It was a small conceptual leap, but measured in terms of twentieth-century consequences it may well be unrivaled. No single discovery had as much impact on the explosion of population growth as Haber's artificial ammonia. There were roughly two billion people alive on the planet when Haber first began his experiments. Today there are 7.7 billion people alive. And yet despite that explosive growth, rates of starvation and chronic malnutrition have plummeted. The mass famines that once killed tens of millions in a year have been entirely eliminated. The Haber-Bosch process—and the subsequent innovations known as the green revolution—led to extraordinary increases in agricultural productivity, breaking through the limits to population that have concerned critics from Malthus to Paul Ehrlich's doomsday predictions in 1968's *The Population Bomb*. Farmland covers roughly 15 percent of the earth's surface today. If crop yields had stayed at 1900 levels, more than half of the ice-free land on the planet today would have to be devoted to farming—much of it with soil that would not support intense agriculture without artificial fertilizers. As in the story of smallpox eradication, global institutions have played a critical role as well. When temporary food shortages arise in conflict zones or sites of natural disasters, organizations like the World Food Programme, the recipient of the 2020 Nobel Peace Prize, intervene to prevent catastrophic famines on the order of the one that erupted in Persia a hundred years ago.

THE BATTLE AGAINST HUNGER and mass starvation carried out in the twentieth century was not exclusively fought in the soil, however. It also involved a controversial revolution in livestock production, what critics now deride as "factory farming." No single animal embodies the vast scale of this revolution more than the chicken. It seems strange to imagine this now—in an era where chicken has become a staple of diets all around the world, amplified by its prominence on American fast-food menus—but until the first decades of the twentieth century, chickens were largely bred for egg production, not for meat. Many households maintained their own coops, and only served chicken at the dinner table when one of the birds was culled because it wasn't producing enough eggs.

One of the initial triggers that transformed the role of chicken in our diet involved a simple typo and an accidental entrepreneur. In the early 1920s in Sussex County, Delaware, a young woman named Cecile Steele had been maintaining a small flock of laying chickens on her family farm, mostly to supply eggs for her family, though she would occasionally sell surplus eggs to bring in extra income. Each spring, she would order fifty additional chicks from a local hatchery. But in the spring of 1923, a mistake at the hatchery added an extra zero to her order; to Steele's surprise, *five hundred* chicks showed up on her doorstep. A less enterprising customer would have simply returned the excess supply, but something about the sight of all those chicks planted an idea in Steele's mind. She stored them in an empty piano box until a lumberman could build a new shed large enough to house them. Steele fattened them up with newly invented feed supplements, and when they reached two pounds, she sold 387 of them for sixty-two cents a pound, making a tidy profit. The next year she deliberately increased the order to one thousand chickens and began scaling up the facilities on the farm. Up until that point, most chicken purchased by restaurants or grocery store chains had been older hens, sold to be used in stews. Steele's poultry was young, which meant the meat was more tender and better suited for frying.

Five years after the fateful delivery of those five hundred chicks, Steele had built out one of the first factory-style chicken farms, raising and selling 26,000 birds in a single year. Within a few years, the number had grown to 250,000. Hundreds of farmers in the region took notice of Steele's success and launched poultry farms that emulated Steele's. They discovered that chicken broilers—as they were called—were more efficient producers of protein than cattle or pigs; they required far less space and they grew to market size in just a matter of weeks, while cattle could take more than a year.

By the 1950s, the poultry industry had discovered that feeding chickens vitamin D supplements—fortified with antibiotics— allowed them to live indoors without exposure to sunlight; before long, industrial-scale coops crowded as many as thirty thousand chickens into wire cages so small that the birds did not have room to spread their wings. The result was a dramatic increase in the efficiency of meat production: you could produce one pound of chicken meat with just two pounds of grain, while a pound of beef required *seven* pounds of grain. That efficiency produced what one writer called "a vast national experiment in supply-side gastro-economics"[10]: markets were flooded with inexpensive chicken and diets quickly adapted. Fast-food chains like Kentucky Fried Chicken (KFC) proliferated. McDonald's added Chicken McNuggets to its worldwide menu in 1983, shortly after concern about the relationship between heart disease and fats caused a special Senate committee on nutrition to recommend that Americans "decrease consumption of meat and increase consumption of poultry and fish." Today the average American eats more than ninety pounds of chicken a year. Industrial poultry production has played a critical role in feeding exploding populations around the world. In 1970, Brazil produced 217 metric tons of broiler meat; today they produce around 13,000 metric tons. Both China and India have seen their chicken meat production grow by more than a factor of ten over the past two decades.[11]

But the scale of this transformation is perhaps best measured by one single data point: the overall population of chickens worldwide. The most numerous wild bird on the planet is the African redbilled *Q. Quelea*, with an estimated population of 1.5 billion. At any given moment, something in the range of twenty-three billion chickens are alive, and human beings consume more than sixty billion chickens each year. (The second number is so much larger because chickens are slaughtered for meat after only a few months of life.) There are now more chickens on earth than all other species of birds combined. The rate of population growth of chickens far exceeds that of humans over the past century. But, of course, the two growth rates are fundamentally linked: we can support seven billion people on the planet now in part because they have 60 billion chickens to eat each year.

The chicken population on earth is so immense, in fact, that scholars now believe that when future archaeologists thousands of years from now dig through the ruins of what some call the Anthropocene age—the age where humans began transforming the planet—they will use the remains of all that poultry as a key marker for the period. No doubt they will encounter other evidence of human culture: nonbiodegradable plastics, buried cities. But the remaining traces of *Homo sapiens* skeletons will be an afterthought to those future archaeologists. The defining biological signature of the period, mummified in landfills all around the world, will be chicken bones.

THE IMPACT OF the agricultural revolutions of the twentieth century—both the increase in soil fertility and the factory farming techniques that brought all those chickens into the world—stagger the mind. Experts believe that these agricultural revolutions doubled the carrying capacity of the planet, which means that without these breakthroughs, half of the 7.7 billion people alive today would never have been born, or would have died of starvation long ago. Countless others would have lived, but at the very floor of their metabolic capacities, barely able to function. Fifty years ago, more than a third of the people living in developing countries were chronically undernourished. Today, just over ten percent of them are.

As Robert Fogel has argued persuasively over the years, increased nutrition can create positive feedback loops of "technophysio evolution": new scientific breakthroughs increase the caloric intake of humans, which gives them more energy for work and economic productivity, which then leads to more innovations that further increase caloric intake. It is no accident that many regions of the world with the most spectacular rates of growth since World War II— many of them in Asia—are places where caloric intake has gone from borderline starvation to levels comparable to modern Europeans.

The escape from hunger is one of the great triumphs of the

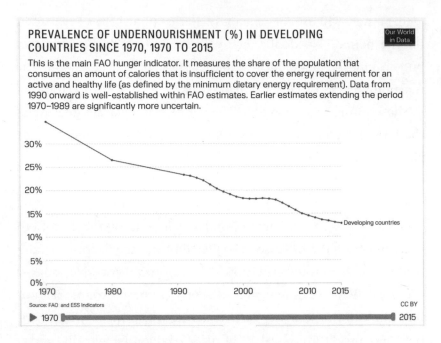

PREVALENCE OF UNDERNOURISHMENT (%) IN DEVELOPING
COUNTRIES SINCE 1970, 1970 TO 2015

Our World in Data

This is the main FAO hunger indicator. It measures the share of the population that
consumes an amount of calories that is insufficient to cover the energy requirement for an
active and healthy life (as defined by the minimum dietary energy requirement). Data from
1990 onward is well-established within FAO estimates. Earlier estimates extending the period
1970–1989 are significantly more uncertain.

Source: FAO and ESS Indicators   CC BY

▶ 1970   2015

Prevalence of undernourishment in developing countries, 1970–2015

twentieth century, but it was not without its costs. The production
of artificial fertilizers consumes as much as 5 percent of the world's
natural gas supplies; artificial fertilizer runoff from farmland has
created massive dead zones in seawater near river deltas, with the
nitrates depriving marine life of sufficient oxygen to survive. As I
write, an area of more than eight thousand square miles in the Gulf
of Mexico is believed to be entirely devoid of life, one of the largest
dead zones ever recorded. A planet carrying twenty-three billion
chickens is also running a massive and unprecedented experiment
in inadvertently breeding new strains of avian flu. The H1N5 virus
that provoked such global panic in 2007 was partially transmitted
by chickens. If another pandemic emerges in the coming years with
even more devastating effects than COVID-19's, the immense pop-

ulation of chickens on earth—and the systems of factory farming that produce them—is likely to be a point of origin for the outbreak.

And even if the 7.7 billion people alive today do not end up contracting new diseases, their existence puts additional strains on the planet, both in terms of environmental destruction and the output of greenhouse gases. We face the global crisis of climate change not just because we adopted an industrial lifestyle, but also because we figured out new techniques to keep people from perishing in mass famines or living at the very edge of starvation. Some of those techniques happened to have some unlikely origins—bat guano and bomb making, an accidental order of chicks—but their ultimate impact is almost beyond comprehension: billions of lives lifted out of hunger and starvation, and a planet struggling to manage the secondary effects of that runaway growth.

CONCLUSION

BHOLA ISLAND, REVISITED

This book began with two simple charts: one that condensed millions of human lives over the past four centuries into a single line, moving up and to the right, and one that tracked the astonishing drop in childhood mortality rates over the last two centuries.

But those charts tell the story of averages, not distributions. How encouraging is the data when we look at the inequalities—the gradients in life expectancy, not the mean? In 1875, just as the great escape was beginning to appear in England's working classes, the gap between the wealthiest British citizens and the rest of the population was a whopping seventeen years. Today the gap still exists, but it is a fraction of its former self: just four years. US health data tells a similar story: the gap between white and African American life expectancies dramatically reduced over the past century, to just under four years. In 1900, just after W. E. B. DuBois first documented the impact of racism on health outcomes, the gap was almost fifteen years.

But arguably the most encouraging trend is the one documented in this chart, zoomed into the last seventy years.

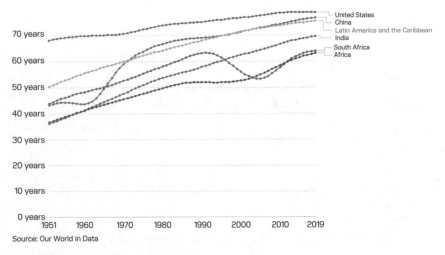

Source: Our World in Data

Life expectancy at birth, 1951–2019

The gap between what we once called industrialized and pre-industrial states, that we now generally describe as the gap between the West and the Global South, has been narrowing for the past thirty years at a rate unheard of in the history of demography. It took Sweden 150 years to reduce childhood mortality rates from 30 percent to under 1 percent. Postwar South Korea pulled off the same feat in just forty-five years. At the end of World War II, life expectancies in India were still trapped at the long ceiling of thirty-five. Today they are above seventy. In 1951, the gap that separated China and the United States was more than twenty years. Today it is just four. Many residents in Western countries think of the past few decades as an age of skyrocketing inequality, and indeed *within* those countries, particularly the United States, the economic outcomes in particular have been winner-take-all affairs. But when you look at the global picture, the image inverts: it is an age of increasing *equality*. The gradients are narrowing.

The reduction in inequality applies to both health outcomes and to income.[1] The Global South is getting wealthier faster than the United States and Britain did in the first wave of industrialization. China— and to a lesser extent, India—are driving those gains, but in the past ten years Africa has shown encouraging resilience as the HIV crisis begins to subside. There was a suspicion held by many that the gradients that separated the West and the Global South were predicated on a kind of planetary zero-sum game, where the affluence and longevity of the West was something that could only be achieved in a global system where half the planet lived in abject poverty. The success of the "developed" nations was dependent on exploiting the resources and labor of the "undeveloped" nations. This may have indeed been part of the dynamic during the peak years of the slave trade and colonialism, but it does not appear to be the case any longer.

The relationship between economic and health progress in the Global South is almost certainly a symbiotic one. In this respect, these less affluent nations are experiencing the same techno-physio evolution that the Nobel laureate Robert Fogel detected in nineteenth-century European society. Removing whole segments of the population from the debilitating state of starvation and chronic illness creates a new labor pool endowed for the first time with enough metabolic energy to contribute to the economy. That in turn raises the standard of living for those communities, improving their health outcomes, and creating more energy that can be applied to income-producing labor.

"If anything sets the twentieth century apart from the past," Fogel observes, "it is this huge increase in the longevity of the lower classes."[2] There are few forms of progress more unambiguous than this extended life. Most of the other hallmarks of our supposed advancement as a society can be gainsaid by reasonable people: Are we *really* better off with supercomputers in our pockets? Is an automobile-based culture—even a safer one—truly an improvement over the walkable communities of pre-twentieth-century civilization? But it is hard to dismiss the gains when your children don't die of smallpox at two or in a car accident at twenty. It is easy, however, to *ignore* those gains, because they have accumulated over the years incrementally, the aggregate effect of countless interventions removing items from Jefferson's "catalogue of evils." This kind of progress is hard to perceive not just because it is slow, but also because by definition it comes in the form of nonevents, deaths that would have happened a century ago that were avoided altogether. Each time we take that antibiotic that kills off a lingering infection or stop short of a car accident because our antilock brakes kick in, we go on with our lives, barely even registering what has just happened. But in an alternate time line without those protections, we might well have ceased to exist.

As momentous and laudable as those interventions have been, they should not be invoked as an excuse for simply sitting back and letting the march of progress continue. The "catalogue of evils" has many pages left in it. A few years ago, the New York University Langone Department of Population Health created an online tool that allows you to compare the average life expectancies between different census tracts in the United States—a kind of digital-age descendant of those life tables William Farr constructed for London, Liverpool, and Surrey.[3] Where I live in Brooklyn, the average life expectancy is eighty-two, slightly higher than the overall US average. But just twenty blocks away, in the poorer, largely African-American neighborhood of Brownsville, the average is seventy-three years. That is the most fundamental form of inequality you can imagine: almost ten years of life that one community gets to enjoy, but their near neighbors do not—inequality that was first mapped out by William Farr and W E. B. Dubois more than a century ago. The COVID-19 pandemic has offered stark new evidence of health inequalities in the United States. In New York, African Americans were twice as likely as white people to die from the disease. In Chicago, African Americans make up 29 percent of the population, but they account for 70 percent of the deaths related to COVID-19. Outside the United States, the 2014 Ebola outbreak in West Africa was a reminder that many of the world's poorest or most war-torn countries remain, in the words of Partners In Health founder Paul Farmer, "clinical deserts"— regions lacking the basic infrastructure of supportive care: ventilators, dialysis machines, blood transfusion capabilities. Martin Luther King Jr. observed in a speech in 1966, "Of all the forms of inequality, injustice in health is the most shocking and the most inhuman." More than a half century later, we are still fighting that injustice.

All of which means that when we think about progress—with health or any other measure—our crucial task is to look at the data from two angles simultaneously: we need to study past trends to learn what has worked for them, and in the process be inspired by their successes. But we also need to keep an eye on all the ways in which the present, given its potential, is underperforming. What technology or intervention currently part of our adjacent possible could reduce mortality further? Yes, my fellow Brooklynites in Brownsville would probably have seen life expectancies ten years shorter back in the 1970s, so on one hand we should celebrate the progress that community has experienced since then. But we should also stay focused on removing the gap—those scandalous ten years of expected life—that currently separates my neighborhood from theirs. It's not enough simply to remind ourselves that progress is possible. It's just as important to figure out what's left to do.

MOST OF THIS BOOK has been devoted to the specific stories behind individual advances in health, mapping the networks that brought them into the world. But what happens when we look at the major breakthroughs as a group? Think back to the ranking of life-saving innovations that we began with:

MILLIONS:

AIDS cocktail

Anesthesia

Angioplasty

Antimalarial drugs

CPR

Insulin

Kidney Dialysis

Oral Rehydration Therapy

Pacemakers

Radiology

Refrigeration

Seat belts

HUNDREDS OF MILLIONS:

Antibiotics

Bifurcated needles

Blood transfusions

Chlorination

Pasteurization

BILLIONS:

Artificial fertilizer

Toilets/Sewers

Vaccines

The most striking thing about this pantheon is how few of them originated in the private sector. In terms of the original break-throughs themselves, few of them emerged exclusively as a proprietary advance created by a for-profit company. One notable exception is the three-point seat belt Nils Bohlin designed for Volvo. But one of the reasons the seat belt succeeded, as we have seen, is that Volvo released it to the world as an open patent. And, of course, most of the automobile industry had to be dragged kicking and screaming into making seat belts a default option on all of its cars. Most of the items in this catalog of blessings emerged outside the private sector:

in academic research (think Alexander Fleming and the Dunn School), or the work of enterprising doctors (think Edward Jenner and the milkmaids), or the desperate innovations of field-workers struggling to hack together solutions in the middle of a crisis (think Dilip Mahalanabis and the Bangladeshi cholera outbreak).

It is true that in many of these cases, private companies were instrumental in amplifying innovations that originally emerged out of public sector exploration. While academic scientists and the US military took penicillin from a mysterious mold to a working drug, Merck and Pfizer helped refine the techniques to mass-produce it; those companies—along with several others—would go on to discover and mass-produce other antibiotics as well in the wake of penicillin. The story of insulin offers a comparable lesson. The drug was first discovered and applied as a treatment for diabetes by a group of scientists at the University of Toronto, and was released to the world under an open patent similar to the one Volvo used with seat belts. But the vast majority of diabetics today use a synthetic insulin that was developed as a partnership between research scientists at the City of Hope National Medical Center and the private drug company Genentech. This is a pattern that seems increasingly commonplace, now that the Big Pharma companies are selling legitimate medicines and not the snake oils of the Parke, Davis catalogs: the core, life-extending idea emerges from some public sector nexus of academic researchers, often influenced by ideas originating in other fields, but the wider adoption of the idea depends on private sector production and distribution platforms.

Even if recent health advances have increasingly relied on public-private partnerships, the fact remains that we owe the vast majority of those extra twenty thousand days of life to nonmarket innovations. In an age that so often conflates innovation with entrepreneurial risk taking and the creative power of the free market, the history of life expectancy offers an important corrective: the most fundamental and inarguable form of progress we have experienced over the past few centuries has not come from big corporations or start-ups. It has come, instead, from activists struggling for reform; from university-based scientists sharing their findings open-source style; and from nonprofit agencies spreading new scientific breakthroughs in low-income countries around the world. The ratio may shift in the coming decades as private sector companies begin to explore new approaches to immunotherapy, or they apply machine learning to drug discovery. But if the history of our doubled life expectancy is any guide, we will always need the public option on our side as well.

WE SHOULD ALSO NOT ignore the less tangible innovations: Farr's mortality reports, Hill's randomized controlled trials. I think of these as belonging to six primary categories:

WAYS OF SEEING. Microscopes and medical imaging technology gave us a direct look at some of the pathogens and rogue cells that were killing us, which helped us dream up new ways of fighting back against them. But so did John Snow's map of the Broad Street outbreak; so did the ring vaccination approach that William Foege invented on the fly in Liberia. Seeing the pattern that outbreaks took geographically—the bird's-eye view—turned out to be just as important as the tight focus of the microscopic lens.

WAYS OF COUNTING. William Farr was trained as a doctor, but the vast majority of lives he helped to save in his career came from the work he did with numbers: tracking and documenting the causal relationship between urban density and mortality rates, compiling data that helped Snow dismantle the miasma theory.

WAYS OF TESTING. You can't put the randomized controlled trial on display in a science museum, but the method gave human beings a superpower as revolutionary as any miracle drug or FMRI machine: the ability to distinguish between a false cure and the real thing. Government regulators were then able to use those RCTs to restrict the market to genuine medicines and force the hucksters out of business. Partly these were breakthroughs driven by new statistical methods, but they also involved invention of new institutions and regulatory bodies like the FDA.

WAYS OF CONNECTING. Widening our networks has not always had a positive effect on life expectancy. Just think of the catastrophic body count from smallpox during the Columbian exchange. But think, too, of the international commingling of ideas that brought Mary Montagu back from Constantinople with the secret of variolation etched on her young son's arm. Or think of the international journey of penicillin. Florey and Heatley administered that first 200 mg of penicillin to Albert Alexander on February 12, 1941. By July, the two of them were on a plane to New York, thanks to the connective link of Warren Weaver and the Rockefeller Foundation. Shortly after that, they were in the cornfields of Iowa, tinkering with vats of fermenting corn steep liquor. Had their idea not been able to migrate at such speed, and with such precision, it's entirely likely that the drug would not have been developed in time to make a difference in the war.

WAYS OF DISCOVERING. The antibiotics revolution began with Fleming's accidental discovery of and the international collaboration to manufacture penicillin, but it eventually needed all those soil samples collected by the military and the drug companies to seek out other molecules that might productively combat deadly bacteria. The R&D labs created by Big Pharma in the twentieth century gave medicine a comparable exploratory power: experimenting with thousands of intriguing compounds, looking for magic bullets in the mix.

WAYS OF AMPLIFYING. Vaccines were functional as a medical intervention in the eighteenth century, but it took evangelists using popular media like Dickens's *Household Words*, "a weekly journal," to bring them to mass awareness. Louis Pasteur had hit upon a reliable scientific technique for ensuring milk safety in the 1860s, but it took Nathan Straus's milk depots and genius for publicity to make pasteurized milk the standard.

A number of factors prevent us from celebrating these kinds of innovation. John Graunt studying the Bills of Mortality by candlelight in the 1660s; Farr dreaming up ways of visualizing the mortality impact of dense urban living; Austin Bradford Hill and Richard Doll grasping the importance of randomization in clinical studies—these are all breakthroughs that revolved around doing new things with data. They didn't produce a shiny new contraption; they didn't generate dynastic wealth for their creators; their effect on everyday life was subtle and indirect. But in the long view, they helped billions of people around the world cheat death, creating platforms that enabled the discovery of countless other more direct health interventions. You can extend life with a miracle drug or a new form of surgery or an fMRI machine. But you can also extend life by crunching the numbers, or making a public stand in support of a new treatment, or creating institutions that allow for new kinds of global collaboration.

The unprecedented speed with which a safe and effective vaccine was developed to fight COVID-19 is a perfect example of how these less tangible meta-innovations can work. Yes, the end result was a material object in the form of an injectable vaccine, but the innovations that made that possible were, in many cases, ones that revolved around new kinds of data collection and sharing. When the SARS-CoV-2 virus first emerged in China in the final weeks of 2019, the organism was identified within a matter of weeks. (By contrast, just four decades ago, at the beginning of the AIDS pandemic, it took three years to identify HIV.) And within days of the coronavirus discovery, the genome of the virus had been sequenced, and that genetic profile had been shared with research labs around the world. That genetic information allowed scientists to build the basic architecture for the COVID-19 vaccine in about forty-eight hours. It was in many ways thanks to the astonishing speed of that initial sharing of information that firms like Moderna and Pfizer were able to ship functional vaccines by the end of 2020, exceeding the expectations of even the most optimistic health officials. Imagine the COVID-19 pandemic, except it takes scientists three years just to identify the virus itself. That would have been our reality had SARS-CoV-2 emerged merely a few decades ago.

WHICH INNOVATIONS WILL SHAPE the next chapter in the story of human life expectancy? In low-income countries, the great nemesis that John Snow first identified almost two centuries ago—

waterborne diseases—is still the second most common cause of death, just behind heart disease. Because these deaths so often occur in young children—while heart disease is a condition that almost always develops much later in life—waterborne illnesses have the biggest impact on overall life expectancy. In communities lacking the resources for building large-scale waste removal infrastructure, like Bazalgette's sewers, the most tantalizing new approach involves reinventing the toilet itself. In 2017, the Bill & Melinda Gates Foundation began testing prototypes in India and South Africa of a new toilet designed to work off-grid—lacking any connection to sewer lines, water pipes, or electricity. The toilet is a closed loop that collects human waste and then burns it as fuel, using the energy created from that process to disinfect water. The operating cost for the device is just a nickel a day.[4]

Another hugely consequential intervention in the low-income countries of the world would be the elimination of malaria. When people are asked about the creatures that evoke the most terror, sharks and snakes usually come to mind, but no multicellular organism has been responsible for more deaths over the course of human history than the mosquito. The World Health Organization (WHO) estimates that each year more than two hundred million people contract the disease—caused by the *Plasmodium* genus of parasite transmitted through the bite of a mosquito—and half a million die from it, most of them young children.[5] But the disease is now heavily concentrated in just a handful of African countries, and deaths have already been significantly reduced thanks to the adoption of insecticide-treated bed nets as well as new antimalarial drugs. Because mosquitoes can travel relatively long distances compared to viruses, the ring vaccination approach that led to the elimination of smallpox cannot be easily applied to the problem of malaria.

Scientists are now exploring a radical new approach to eradication, relying on gene drive technology, an emergent form of genetic engineering that compels a trait to spread through the population by altering the odds that a specific allele will be passed on to its offspring. (In a normal organism, each allele has a 50 percent chance of being passed down; increase those odds and the trait in question will rapidly spread through the population, particularly with a creature that has a reproductive cycle measured in days.) The most controversial approach would aim to reduce the mosquito population by passing down a trait that induced infertility in the next generation, with the aim of dramatically reducing, if not eliminating, the overall number of mosquitoes in the wild.[6] Another approach spreads a mutation that makes the mosquitoes resistant to the *Plasmodium* parasite.

As infectious diseases have grown less common in high-income countries, the leading causes of death have shifted toward the chronic illnesses of an aging population: heart disease, Alzheimer's. In 1900, cancer was only the eighth leading cause of death, far behind ailments like gastrointestinal infections and tuberculosis that no longer make the top twenty. At some point in the next few years, cancer will become the leading cause of death in the United States, unseating heart disease for the first time. And yet the last few years have been the most exciting time in the long and frustrating history of the war on cancer, thanks to the emergence of new immunotherapy techniques.

Unlike the great killers of the nineteenth century—cholera or smallpox—cancer does not originate with some external organism trespassing into your body. Though some cancers do seem to have viral triggers, cancer cells are your cells. They have not been hijacked by some invader to serve its distinct evolutionary aims. What cancer does—reproduce itself through cell division—is part of the life cycle of every cell. It's just that the cancerous cells have gone rogue by refusing to *stop* dividing, thanks to some scrambled code in their genetic instructions. We've known about the runaway growth that those stubborn dividing cells produce for almost a century. What we didn't realize was how common it was for cells to shift into that self-replicating-at-all-costs mode. What we call cancer is happening all the time in our bodies, but modern immunology has made it clear that those acts of insubordination are constantly being shut down by the first responders of the immune system. The vast majority of the time it works: a cell refuses to die, and the immune system swarms into place to ensure that it follows orders.

But every now and then, a cell manages to foil the T cells of the immune system by releasing a signal that causes them to withdraw immediately. Cells often grow at cancer-like rates when recovering from a wound: tissue that needs repairing forces cells to divide at a faster clip, for longer stretches of time than in ordinary conditions. The immune system lets that growth happen because the cells release a signal that activates a molecule known as CTLA-4 on the T cells.[7] By activating CTLA-4, the cancer cells are effectively transmitting a message to the antibodies that reads: "I'm just replicating normally here, growing back the injured tissue; no need to shut me down." Deliberately subverting the meaning of that signal was, ultimately, cancer's greatest trick. In a meaningful sense, cancer kills us because our cells learn how to lie.

Despite the intimate connection we have as organisms to the cancers that grow inside of us—that are indeed part of us—we have from the beginning treated tumors as invaders that need to be obliterated at all cost. First we cut them out barbarically, with no knowledge of the germ theory; then we developed more sanitary surgical techniques; then we began bombing them with chemotherapies and radiation. The radical promise of immunotherapy is that we are using some of the most advanced biochemistry on the planet to give the immune system just enough help that it can regulate the cancer itself, using its far more precise tools.

How do immunotherapies pull that off? They scramble the CTLA-4 signal. The malignant cells try to keep dividing, but the T cells never pick up their "don't mind me" message, and so they swoop in and take the rogue cells offline, the way it's supposed to happen. Philosophically, the immunotherapy approach to cancer is as different from radiation/chemo treatments as those treatments were from the barber-surgeons cutting the tumors out without access to anesthesia. Why subject a sick person to dangerous levels of cell-destroying radiation when you can simply allow the body's natural defense system to do the work for you?

There is something about immunotherapy that suggests a closing of a circle. The first great advance in the story of our doubled life was built upon a similar foundation, despite the fact that biochemistry barely existed as a science at the time. Vaccines—and variolation before them—worked via a comparable cellular magic trick: forcing the immune system to manufacture new antibodies to fight off the threat. Antibiotics, once they got into circulation, did the dirty work themselves: invading bacteria died from direct contact with compounds like penicillin that had been introduced into our bloodstream. But vaccines and immunotherapies rely on a different circuit: they don't drop bombs from the outside; they arm our existing defenses. This may well be the future of medicine: the miracle drugs are increasingly designed to let the body heal itself.

. . .

WHAT ABOUT THE META-INNOVATIONS? Is there a methodological breakthrough on the horizon as significant as Bradford Hill's
RCT or William Farr's mortality reports? Some of the most promising new ideas have been developed or accelerated by the COVID-19
crisis, which prompted multiple new experiments in data gathering
and analysis—experiments that likely saved thousands of lives during
the pandemic. And they may well prevent future pandemics from
developing in the first place.

As unlikely as it might sound, given the existence of organizations like the CDC or the WHO, in the early days of the coronavirus's spread, no single data repository existed where information
about all the known cases could be accessed and analyzed by public
health officials and researchers. But also in the early days of the outbreak, an ad hoc organization of academics around the world formed
to create a twenty-first-century equivalent of Farr's mortality reports:
a single open-source archive of every recorded COVID-19 case anywhere in the world. By early February, the Open COVID-19 Data
Working Group, as it came to be known, had assembled detailed
records for ten thousand cases.[8] By the summer of 2020, an informal
network of hundreds of volunteers has assembled records for more
than a million cases in 142 countries around the world. It may well
be the single most accurate portrait of the virus's spread through the
human population in existence.

Of course, the greatest value in that kind of data set lies in the
clues it can give us about the future path of the disease and how
that path can potentially be interrupted. But again, the work of
building those models has entirely taken the form of impromptu
efforts organized at a handful of academic institutions around the

world. The Johns Hopkins University epidemiologist Caitlin Rivers has argued that the coronavirus pandemic makes it clear that one crucial innovation we need is a new kind of institution, what Rivers called a center for epidemic forecasting. Yet forecasts are only as good as the underlying data that support them, and in the case of disease outbreaks, most of the data collection—even in comprehensive archives like the one assembled by the Open COVID-19 Data Working Group—suffers from a crucial liability: The information is captured too late. Numbers of hospitalizations or deaths are vital statistics to be sure, but they are tracking the end stages in the path of a disease. In the case of COVID-19, by the time the average person makes it to a hospital, about ten days have passed since initial contact with the virus.

With a disease like COVID-19, where presymptomatic and asymptomatic carriers are capable of spreading the virus, the lag in reporting can make the difference between a runaway outbreak and effective containment. A typical case of COVID-19 that ends in a death follows this time line, which can stretch to thirty days or more:

**Infection → Incubation → Presymptomatic spread →
Symptoms and spread → Doctor's visit → Hospitalization →
Intensive care → Death**

In the standard regime, even in the best-case scenario, data collection doesn't begin until day ten, during the doctor's visit. COVID-19 has prompted an inspiring scramble of experiments designed to move the data gathering earlier on the time line. Some of them involve what is called sentinel surveillance. In William Farr's mortality reports, or John Snow's maps of the Broad Street cholera outbreak, the data collected lay at the far right of the epidemic time line, given that they were both tracking deaths. Today we have systems in place to capture data from the middle of the time line, when someone is sick enough to show up at a clinic to be tested or be admitted to a hospital. But sentinel surveillance captures an earlier phase, by testing a representative sample of the general public *before* they develop symptoms. One example of this approach is the Seattle Flu Study, an initiative that began in 2019, which set up testing kiosks, analyzed samples from hospitals, and distributed home nasal swabs to a broad section of the city's population, asking them to send in samples if they developed symptoms of respiratory infection. Tellingly, the program was the first to detect community transmission of SARS-CoV-2 in the United States.

Technology can also help move the time line to the left. The San Francisco-based start-up Kinsa has been selling an internet-connected thermometer since 2014. From the consumer's point of view, the interaction with Kinsa's thermometer is straightforward enough, but behind the scenes the device sends anonymous, geo-located information about the results to Kinsa's servers. That new data stream enables the company to maintain what it calls health weather maps for the entire country, with real-time data on atypical fevers reported down to the level of individual counties.[9]

Starting on March 4, 2020, Kinsa's charts began tracking a statistically meaningful increase in the number of fevers in New York nineteen days before the city went into a full lockdown. (The first case in the city was reported on March 1.) By March 10, the number of people registering an elevated temperature in Brooklyn was 50 percent higher than normal, suggesting that the virus was already rampant throughout the five boroughs, even though the official case load was still less than two hundred.

The most radical technique for shifting the data-collection time line to the left—but the one that might offer the most significant protection against future epidemics—involves cutting people out of the equation altogether. The underlying data that allowed William Farr to draw the first epidemic curve back in 1840 was, understandably, limited to patterns of life and death in the human population. Sentinel surveillance allows us to pick up signals earlier in the cycle by detecting symptoms before people make contact with the health system. But for many of the most terrifying diseases that have emerged in the past few decades, the initial human cases showed up in the middle of a much longer time line, one that dates back to the point where the virus jumped from animals to humans. The epidemiologist Larry Brilliant, who played a key role in the eradication of smallpox during the 1970s, has argued that the most powerful way to move the time line to the left is through animal surveillance—building new systems designed to track outbreaks of disease in factory farms around the world, mostly among those sixty billion chickens that now so dramatically outnumber us.[10]

The promise of applying William Farr's vital statistics to the realm of animal diseases is a simple one: you can stop an emerging zoonotic disease before it makes the jump from animal to human. Animal surveillance could ward off the potential pandemic that experts have historically worried about the most: an influenza outbreak along the lines of the 1918 avian flu, a terrifying unintended consequence of those sixty billion chickens. Public health data began with that most elemental form of accounting: how many people died on this day in this place. But during an epidemic, from the perspective of vital statistics, a human death tells the story of an infection that happened in the past. A hundred dead chickens, on the other hand, could tell the story of a future infection—and maybe even stop it from emerging at all.

For more than a decade, the United States government funded a program that performed exactly this sort of animal surveillance—a program called Predict that collected more than 100,000 biological samples from animals all around the world, discovering more than a thousand new viruses in the process. Despite the fact that it cost a mere $200 million to run over that period—a rounding error in the federal budget—the Trump administration shut down Predict in the fall of 2019, only a few weeks before reports began surfacing of an alarming new viral outbreak in Wuhan, China.

AS IS SO OFTEN the case in the history of human health, some of the most meaningful advances that will shape the next few decades of longevity are likely to originate in seemingly distant fields. In the nineteenth century, one of those unlikely links came from soil science. In the twenty-first century, a comparable life-extending revolution may well emerge from the study of computer games.

In early December of 2017, the DeepMind subsidiary of Google published a research paper documenting the progress it had made with its cutting-edge machine-learning program called AlphaZero.[11] DeepMind had been founded in London seven years before by a polymath named Demis Hassabis, who had spent his twenties oscillating between studying cognitive neuroscience and designing video games, while playing world-class chess on the side. DeepMind spent the first few years of its existence as a start-up that trained algorithms to play video games, slowly climbing the tree of gaming complexity: from Pong to Space Invaders to Q*bert. The simplicity of these early games might make DeepMind's achievements seem less impressive; after all, computers had been regularly defeating world champions at much more challenging games like chess for more than a decade. But Hassabis and his team were working with one critical limitation: they weren't giving their algorithms any cheat sheets. The Deep Blue chess computer that famously defeated Garry Kasparov in 1996 had been equipped with an immense database of previous games, with a library of moves programmed by human grandmasters. It was an amalgamation of human knowledge of chess strategy combined with the brute force computing power that enabled it to draw on that database to calculate potential moves and their effect at superhuman speed. DeepMind's algorithm, on the other hand, came to its games in a state of complete ignorance, with zero information about strategy. It relied on a novel approach to AI known as Q-learning, also known as deep reinforcement learning. The approach is considered "model free" in that the algorithm does not have a preexisting model of the system—in DeepMind's case, the game—it is trying to learn. Instead, it learns from the bottom up, through a near-endless series of iterations, experimenting with billions of different strategies. Hassabis called it tabula rasa reinforcement.

Later in their research, DeepMind began developing AlphaZero with a slightly different approach: the algorithm would learn how to win at board games like Go or Chess by playing games against itself. AlphaZero would start with only basic information about the rules: how pawns can only move one square at a time, while bishops can only move diagonally, and so on. Beyond that bare bones knowledge, AlphaZero came to its first game of chess as a complete blank slate. Of course, the player at the other side of the virtual chessboard was equally ignorant, given that it was a duplicate version of the algorithm. Unsurprisingly, their first games were astonishingly bad. A third grader who had just joined the chess club could have defeated them. But just nine hours later, AlphaZero had become the most advanced chess player on the planet. That seems like a preposterously short amount of time to accumulate so much knowledge, but the algorithms had been busy during those nine hours, playing *forty-four million* games of chess during a single workday. By comparison, the average human grandmaster might play somewhere on the order of a hundred thousand games over the course of his or her life.

Intriguingly, the style of play that AlphaZero evolved over those nine hours had an unusual aggressiveness compared to human grandmasters. In a subsequent paper, DeepMind analyzed a stretch of the training process where the algorithm had independently hit upon a set of strategies long employed by high-ranking players; after deploying them in a few hundred thousand games, AlphaZero ditched them for a more effective approach. (Musing on the achievement in the *New Yorker*, the writer and programmer James Somers observed, "It is odd and a little unsettling to see humanity's best ideas trundled over on the way to something better."[17]) Grandmasters had spent centuries slowly assembling the expertise required to perceive those deep patterns of strategy; AlphaZero bootstrapped its way to them in a few hours—and then left them behind.

I suspect fifty years from now, we will look back on those forty-four million games as a milestone in the history of human health every bit as significant as the day the handle was removed from the Broad Street pump or the morning Alexander Fleming returned from his vacation to discover that moldy petri dish by the window. The ability to play chess is only a small subset of human intelligence; the fact that DeepMind can breed grandmasters in an afternoon says little about their ability to create machines that have a general intelligence to rival that of *Homo sapiens*. And yet the kind of adversarial, open-ended learning that AlphaGo displayed is particularly well suited to the biochemistry of health. (As it happens, it is not dissimilar from the way the immune system learns to attack pathogens that it has never experienced before.) Instead of tinkering with new chess strategies, the algorithm will someday explore novel compounds that could be used to destroy deadly viruses, or switch off the runaway growth of cancer cells, or repair the damaged neurons of Alzheimer's. Tellingly, the first product released by DeepMind that was not devoted to gameplay was an algorithm announced in 2018 called AlphaFold, designed to predict the 3D structure of proteins based on genetic sequences—a process that is critical for understanding diseases like Parkinson's or cystic fibrosis that result from "misfolded" proteins—as well as for designing new drugs to combat a much wider range of illnesses.

In a sense, what algorithms like AlphaFold and its descendants may end up doing is the digital equivalent of all those military servicemen bringing back soil samples from around the globe in the middle of World War II, or Mary Hunt browsing the produce aisles of the Peoria markets. Instead of searching for promising microbes in mine shafts and moldy cantaloupes, the software will explore billions of combinations, stringing together virtual amino acids to make the complex 3D shapes that govern our health on a cellular level. It will be a discovery mechanism, pushing the boundaries of the adjacent possible by playing millions of simulating "games" against a simulated pathogen, dreaming up promising new protein structures that might outfox the enemy.

If tomorrow's deep learning algorithms do indeed end up playing the role of Mary Hunt and her moldy cantaloupe, it will be a particularly timely breakthrough. Almost every single antibiotic on the market today was discovered before 1960, during the great flurry of activity that followed the development of penicillin. Over-prescription and the adoption of antibiotics in industrial livestock diets has led to a troubling resistance in recent years, as bacteria evolve new strategies for evading or counteracting these former miracle drugs.[13] Algorithms like the ones DeepMind has pioneered may indeed increase the speed and range of the drug discovery process, enabling us to engineer new compounds at a rate faster than the bacteria can evolve resistance. But antibiotic innovation has not just stalled because we have not had the tools for making new discoveries. It has also stalled because Big Pharma has lost interest in the field. Expensive drugs to treat cardiovascular disease and cancer are where the money is. Cheaper drugs like penicillin and its descendants—which are only taken in small doses—don't move the needle economically for these for-profit firms; in all likelihood, developing a new antibiotic from scratch that dramatically outperformed the others enough to raise prices would cost tens of billions of dollars.

This potential market failure has led some figures to call for the formation of a new kind of institution: a global NGO of sorts that would produce and distribute existing antibiotics, and actively fund the development of new variants, perhaps using some of the technologies underway at DeepMind. (Its nearest precedent, fittingly, is the hybrid network that brought penicillin into the world: the Dunn School, the Rockefeller Foundation, the US Department of Agriculture [DOA], the US armed forces, and a handful of private sector actors like Merck and Pfizer.) Some of that mission has already been funded at significant scale by organizations like the Wellcome Trust, which has spent more than $600 million to date supporting antibiotic research. But a single-purpose entity—endowed with tens of billions of dollars of resources, drawing on collaborative networks that extend around the world—could potentially provide the next step in our collective relationship with life's smallest organisms. DeepMind gives us new ways of exploring and seeing: crunching the data that describe those amino acid chains, visualizing the myriad ways they could combine with the protein folds of the bacterium. But a global nonprofit devoted exclusively to antibiotics would give us a genuinely new way of amplifying.

THEN THERE IS THE MATTER of the blind spots. If history is any guide, the medical establishment is no doubt walking around with some kind of widely held consensus that will turn out to be fundamentally wrong a few decades from now, just as the miasma theory evaporated under the light of Snow's map and Koch's microscopes. What core part of today's health orthodoxy will our grandchildren be baffled by?

The most controversial answer to that question is one that has been bubbling up from the fringes of the medical establishment and the transhumanists of Silicon Valley. Our biggest blind spot, they argue, is our outdated belief that life has to end. What if aging itself could be removed from the "catalogue of evils"?

Some of this enthusiasm stems from a simple statistical misperception, perhaps the most common of all the ways our brains have trouble with probabilities. If you do not factor in the importance of childhood mortality declines, it seems as if the human race is on a clear path to near immortality: a hundred years ago, the average person used to die at forty, and now the average person lives to eighty. Just keep those trends going for another few decades and we will reach a kind of demographic escape velocity as a species. But, of course, the mean life expectancy is misleading: the most dramatic change compared to a century ago is not people living into their hundreds, but rather how much more likely we are to survive childhood.

The serious researchers who are investigating the immortality question understand these demographic issues, of course. Their belief in the possibility of reversing the clock of aging stems not from the achievements of the past century but from new understanding of what is called the epigenome, the system of chemical agents that activate DNA and regulate its expression. Every cell in your body contains in its genetic code the complete instruction set for building all the different types of cells that make up a human being: liver cells, blood cells, neurons, and so on. But a liver cell only expresses the parts of the instruction set that are relevant to the production of liver cells, because the epigenome has regulated that expression. Scientists now believe that the aging process itself is the result of specific epigenetic instructions. Aging, in this scenario, is not just a third-law-of-thermodynamics inevitability, the unavoidable decline of wear and tear. Human beings in their twenties show almost no sign of age-related decay because their cells are still under orders to keep themselves in full working order. But for some reason, after we turn thirty, those self-repair instructions get less strict. From evolution's perspective, aging may be a feature, not a bug: repair the body's cells long enough to make it through the procreative years, and then switch off the maintenance routines so that the next generation has its turn. Or perhaps natural selection simply failed to come up with a way to keep the self-repair cycle looping. Either way, we don't die of old age just because things fall apart. We die because our epigenome has decided that we aren't worth the trouble anymore.

But what if we could flip that switch, the way immunotherapy blocks the CTLA-4 signal? About a decade ago, a Stanford genetics professor named Howard Chang discovered that the release of a protein called NF-kB triggers the aging process in skin tissue cells; inhibiting the protein in older mice caused their skin to look noticeably younger.[14] The discovery suggested a profound possibility. The human body is constantly generating new epidermal cells; the average life span of a skin cell is only two to three weeks. And yet the epidermal cells of an octogenarian don't look like the cells of a two-week-old baby. On a cellular level, the new skin on an older person comes into the world pre-aged. But one crucial biological event does reset the clock: the creation of a fertilized egg. When two forty-year-olds have a child, their respective sperm and egg cells display the distinctive signs of aging, the result of epigenetic signals that have turned off their capacity for self-repair. But the zygote they produce displays none of those signs of age. Something in the reproductive process is capable of arresting the steady decay of aging, making new cells in an old body.

Right around the time that Howard Chang was injecting his mice with NF-kB inhibitors, a Japanese biologist named Shinya Yamanaka published a groundbreaking study that documented the four crucial genes responsible for the clock-resetting of the newly fertilized egg. In late 2016, a geneticist at the Salk Institute named Juan Carlos Izpisua Belmonte announced that he and his colleagues had engineered mice with an extra set of Yamanaka's four genes. Belmonte created a kind of external epigenome to activate those genes: the Yamanaka factors, as they are called, were only switched on when the mice drank a drug that Belmonte would put in their drinking water twice a week.[15] Earlier experiments where the Yamanaka factors were constantly operational killed the mice, but for some reason, only triggering the self-repair cycle occasionally produced far better results. The engineered mice lived 30 percent longer than the control group. Their lives were extended not by defeating a chronic disease or killing off a bacterial invader, but rather by a new kind of intervention: slowing the aging process itself.

RESETTING OUR CELLULAR CLOCKS may never be possible, or it may be a biotechnology that's still hundreds of years away, given how complex the aging process likely is. But say, for the sake of argument, that advocates like Aubrey de Grey and the other transhumanists are right; say we are on the cusp of raising the ceiling of life expectancy even further and faster than we did over the past century. What would the society-wide implications of such a development be? We have some experience in this department, thanks to those extra twenty thousand days we've already gained. We have seen how mortality reductions can lead to explosive population growth, even as birth rates decline. We have seen the damage those growth rates can have on the earth's environment. The ecosystems of this planet have been coevolving with humans for millions of years, but the vast majority of the time the total population of *Homo sapiens* numbered in the hundreds of thousands. There were only five hundred million people on the planet when John Graunt first starting counting up the deaths in the 1660s; just two billion when the Great Influenza first struck. Today there are almost eight billion.

Imagine what happens to that number if people start opting to freeze their biological clock at the age of twenty-five and live for centuries.

Almost certainly, the initial products that go on sale offering an RCT-tested cure for aging will be costly, to say the least. Peter Thiel and Ray Kurzweil will be first in line, but the treatments will be far too expensive for even a middle-class budget in the United States, much less one in Nigeria. After a century of declining inequality, a new gradient will open up in the mortality tables: between the rich and the poor, the immortals and the mortals. That alone suggests profound ethical questions: Is it right to allow some people to live forever, while condemning others to death and the slow decline of aging, based solely on how much money they have in the bank? Is it right to offer that choice only to the wealthiest people in the wealthiest nations?

Then there is the question of the impact on global population. Jumping from two billion to eight billion in just a century creates a terrifying line graph if you assume the parabola continues to climb. But there is good reason to believe that global population numbers will stabilize over the next few decades, as the societies in the Global South go through the same "demographic transition" that the first industrialized countries went through in the 1800s. This pattern has been observed again and again around the world since it originally appeared in Europe. It follows a predictable sequence: reductions in childhood mortality swell the population, with millions of babies who would previously have died before reaching adolescence now living long enough to procreate. Families continue conceiving children at the same rate because the mortality reductions take time to become apparent, time to be integrated into the social norms of society. By the time they've realized that all their offspring are going to survive into adulthood, it's too late to shift strategies. And so there's a lag during which population swells. But eventually modernization brings more women into the labor force, and often into dense cities, making them less interested in having large families. The increased access to education and birth control that often accompanies industrialization gives them new tools to reduce the number of pregnancies. In many of the societies that first went through the "demographic transition," birth rates have dropped below replacement levels, with an average family size of less than 2.1 children. Assuming this pattern will hold for the Global South—it certainly appears to be holding for China, in part thanks to government regulations that are understandably abhorrent to many Westerners—global population growth should level out circa 2080, somewhere north of ten billion. After that, our footprint will finally start shrinking again.

But not if we stop aging.

Perhaps there will be similar adjustments in childbearing practices as people grow to accept the premise that they have centuries to live, not decades. There are three primary metrics that govern the relationship between life expectancy and overall population: birth rate, death rate, and average age of parents when they have their first child. A society living longer and having more babies can keep population in check by postponing the average age of becoming a parent. If the average person lives to seventy, and the average parent has the first child at twenty-five, there will be a good number of grandparents in the world, and quite a few great-grandparents. All those generations coexisting add up in terms of the overall numbers. But a society with a life expectancy of seventy, where most people wait until they're forty to have children, will have far fewer grandparents and great-grandparents. Perhaps the possibility of living to two hundred with your body permanently in the form of a twentysomething adult will cause a radical shift in the way people think about becoming a parent. When I was born, the average first-time mother was just over twenty years old; today that number is approaching thirty. Perhaps the immortals will get an entire first pass at a childless career before they decide to settle down and have kids at sixty-five. That might stabilize the growth for a while, but eventually the numbers would catch up to us.

But wherever you happen to land on these ethical dilemmas and back-of-the-envelope forecasts, one thing is undeniable: ending the aging process would be the single most important thing that has ever happened to our species. Living in a world where death was, for all practical purposes, optional would change everything. It would pose enormous new threats to our ability to live within the carrying capacity of the planet. It would challenge many of the central precepts of the world's religions and introduce pernicious new forms of inequality. But at the same time, it would remove the most intransigent item in the "catalogue of evils" and spare billions of people the tragedy of watching their parents, partners, and other loved ones dying—not to mention enduring the pains and indignities of growing old.

A change that profound warrants deliberation. Polls show that most people do not want radically extended life spans. Instead, they want longer "health spans"—the period of time during which you are fundamentally unimpaired by any disease or injury—followed by a quick and painless death. Most people would rather live to one hundred with a sound mind and a functioning body and then drop dead, versus living for centuries.[16] And yet the immortality research is charging ahead, funded by the tech billionaires and prestigious institutions like the Salk Institute. If it is indeed within the adjacent possible for us to reset our cellular clocks, to live life indefinitely as a twenty-five year old, will we just flip that switch as a species without any formal debate? Who should decide whether we take that momentous step? Surely the choice cannot be made exclusively by the people wealthy enough to fund the research. Ending the aging process will require advances in epigenetics and gene editing and a thousand other subdisciplines. But it may also force us to invent new kinds of institutions as well, a kind of global regulatory body that could help us navigate a choice this complex. When Frances Oldham arrived at the University of Chicago as a twenty-two year old, we hadn't yet invented a regulatory agency capable of protecting us from medicines that were accidentally killing people. We may need to invent a comparable institution to help us come to terms with medicines that eliminate death altogether.

There is also the possibility that we are worrying about the wrong problem. A century of rising life expectancies has made that upward march seem almost inevitable: the Moore's Law of public health. But what if those extra twenty thousand days turn out to be an anomaly? In the United States, for the first time since the end of the Spanish flu, average life expectancy has decreased for three straight years. As I write, the COVID-19 pandemic is still encircling the globe. With global temperatures rising and the population explosion continuing until at least 2080, is it possible that the trends in aging could reverse over the next century? Could the great escape be brought back to earth?

IN 1927, a chiropractor named Don Dickson decided to investigate the strange mounds of earth dotted across the landscape of his family farm in central Illinois. It didn't take long for Dickson to realize that he was digging out a significant archaeological site. His own explorations uncovered hundreds of Native American skeletons, buried by the indigenous societies of the Illinois River Valley centuries ago in ceremonial mounds. Dickson did his best to keep the skeletons in place, erected a tent over the excavation, and began selling tickets to what was effectively a pop-up museum. Eventually a traditional visitor center was constructed at the site, and today Dickson Mounds is a member of the Illinois State Museum system, though the skeletons have been removed from the exhibits out of respect for Native American values.

The Dickson Mounds complex turned out to be of great interest to archaeologists—and demographers—for reasons similar to those that drew Nancy Howell to visit the !Kung people back in the late 1960s. The earliest burial sites on Don Dickson's farm—dating back about a thousand years—were dug out by hunter-gatherers in the Illinois River Valley. Because the skeletons were relatively well preserved, paleopathologists could examine them for signs of illness and malnutrition, and construct life tables for the community based on rough estimates of the age of death for each skeleton on the site. The result of that investigation painted a picture of a society similar to what Howell found among the present-day cultures of the !Kung: mean life expectancy was twenty-six, just below the long ceiling; infant and childhood mortality was just over 30 percent. Fourteen percent of the community lived past fifty.

Dickson Mounds offered more than just a snapshot of hunter-gatherer health outcomes, however. The site told a story of change as well. Sometime around AD 1150, the Native Americans in the area shifted from their hunter-gatherer roots and adopted agriculture for the first time, primarily in the form of intensive maize farming. They continued the agricultural lifestyle for another few centuries, until something ended their practice of burying their dead in that region. The switch to farming left indelible marks in the skeletons of the Native Americans who experienced the transition: enamel defects in teeth, which signal chronic malnutrition; bones malformed by iron-deficiency anemia; degenerative spinal conditions that likely were the result of increased hard labor. The life tables told an equally grim story. Mean expected life at birth dropped seven years, to just nineteen. Childhood mortality rates were above 50 percent. Only 5 percent of the population survived to the age of fifty.[17] The adoption of the agricultural way of life was as devastating to the Native American communities as industrialization had been to the families living in Liverpool when William Farr built his first life tables.

The pattern of lives and death made visible through the study of Dickson Mounds has since been replicated around the world by paleopathologists studying the historic transition to agriculture. Again and again, we see mortality rates skyrocket thanks to decreased nutrition, increased infectious disease, and backbreaking labor. Most agricultural societies appear to have taken thousands of years to return to the life expectancies and childhood mortality rates enjoyed by hunter-gatherers. We have a romantic, salt-of-the-earth affection for farming today, but its original appearance as a mode of economic production was every bit as catastrophic as the factories of Northern England were in the early 1800s. Because agricultural societies both reduced life expectancies and introduced new forms of economic inequality, Jared Diamond has called the adoption of farming "the worst mistake in the history of the human race."[18]

The story of grim decline that the Dickson Mounds revealed may seem very far from our current situation, standing as we are at the end of a century of miraculous progress in human health— progress not just measured in the lives and deaths of the most advanced societies but on a truly global scale. But the lesions and broken bones of those skeletons are a reminder that the upward parabola of the great escape is not an inevitability. Earlier societies made collective choices about how they should be organized that caused their lives to be shortened, not extended, creating downward spirals that lasted for millennia. To be sure, there are good reasons to believe that we can avoid another retreat like the one our ancestors experienced at the dawn of the agricultural age. We saw life expectancies plummet at the dawn of industrialization too; and yet the fact that we could see those patterns—in the mortality reports, and the life tables, in the maps of the Broad Street outbreak—suggested strategies for struggle and reform and new innovations that ultimately reversed that decline, in just a generation or two. We have far more powerful tools at our disposal today.

The COVID-19 pandemic witnessed such a tiny fraction of the mortality experienced during the Great Influenza in part because we have scientific and public health expertise that the world lacked a hundred years ago. Scientists were able to identify and sequence the genome of the SARS-CoV-2 virus using tools that would have seemed like magic to the scientists and doctors fighting the 1918 outbreak; the internet enabled them to share that information at the speed of light. When the first vaccines entered Phase 1 trials in March of 2020, drug companies could analyze the results using the statistical techniques that Austin Bradford Hill had pioneered in the 1940s. Machine learning algorithms scoured vast databases of information looking for potential drug combinations that might treat COVID. Epidemiologists were able to build sophisticated models to project the path of the outbreak, convincing authorities that lockdown strategies to flatten the curve were necessary. Almost none of those resources were available to the doctors and public health authorities battling the Spanish flu a hundred years ago. The cost of the COVID-19 pandemic—in lives lost, in economic disruption—was immense, to be sure. And countless mistakes were made, in underestimating the scale of the threat in the early days of the outbreak and in failing to adopt simple public health interventions like mask wearing. But millions more would have perished without the defenses that were ultimately put in place.

It is possible that in the future a virus more lethal than SARS-CoV-2 will outfox our defenses and create a pandemic on the scale of 1918; or perhaps some rogue technology will kill enough people to reverse the great escape. But I suspect the greatest threat to the twenty thousand days of extra life that we have fought so hard—on so many fronts—to achieve is one that, paradoxically, was made possible by that very same triumph. If a hundred years from now life expectancy has declined, the most likely culprit will be the environmental impact of ten billion people living in industrialized societies. We have astonishing tools to perceive global warming and its real and potential impact—thanks to many of the same kind of multidisciplinary, public sector networks that drove life expectancy upward—but we do not yet seem to have the willpower or the institutions to reduce the greenhouse gases in our environment. Extending our lives gave us the climate crisis. Perhaps the climate crisis will ultimately trigger a reversion to the mean.

No place on earth embodies that history and that potential future more poignantly than Bhola Island, Bangladesh. Four decades ago, it was the site of humanity's most extraordinary achievement in the realm of public health: the elimination of smallpox, realizing the dream that Jefferson had envisioned almost two centuries before. But in the years that followed smallpox eradication, the island was subjected to a series of devastating floods; almost half a million people have been displaced from the region since Rahima Banu Begum contracted smallpox there. Today large stretches of Bhola Island have been permanently lost to the rising sea waters caused by global warming. The entire island may have disappeared from the map of the world by the time our children and grandchildren celebrate the centennial of smallpox eradication in 2079. What will their life tables look like then? Will the forces that drove so much positive change over the past century continue to propel the great escape? Will smallpox turn out to be just the first in a long line of threats—polio, malaria, influenza—removed from the "catalogue of evils"? Will the figurative rising tide of egalitarian public health continue to lift all the boats? Or will those momentous achievements—all that unexpected life—be washed away by an actual tide?

ACKNOWLEDGMENTS

I suppose it would be appropriate for any book about the miracle of being alive to be dedicated to the author's mother, but in this specific case, the debt of gratitude has extra significance. For almost half a century my mother has been an inspirational agent of change in creating more equitable, more humane healthcare experiences and outcomes for patients around the world. Thanks to her, I was raised to appreciate from an early age the vital role that healthcare professionals play in society, and to recognize that positive change in the world of health was not just the result of scientific or technical advances, but also the result of activism and advocacy, often by patients and their families themselves. The emphasis in this book on the role of non-specialists and social movements in extending the human life span comes in part from the research I've done over the years on the history of health and medicine, dating back to my book *The Ghost*

Map. But the truth is it also comes from a lifetime of watching my mother at work.

Two other inspirational figures deserve a special mention with this book. My friend and mentor in all matters epidemiological, Larry Brilliant, was an early sounding board for the ideas here, and a valued collaborator on the television side of the project. (Thanks to Mark Buell and Guy Lampard for introducing me to Larry, and for joining in some of those conversations along the way.) Thanks as well to my producing partner, Jane Root, for believing in the potential of this project as a television series, and keeping the faith through a hundred near-death experiences on the way to getting it made. I'm also grateful for the stellar team that Jane has put together at Nutopia, who were instrumental both in developing the ideas in the book and the series, and in tackling the enormous logistical challenge of making a television show in the age of COVID: Fiona Caldwell, Nicola Moody, Simon Willgoss, Carl Griffin, Helena Tait, Tristan Quinn, Duncan Singh, Helen Sage, David Alvarado, Jason Sussberg, and Jen Beamish.

More than any of my projects, this book has benefited from an immense number of conversations with experts from the many disciplines and historical periods that it covers: Bruce Gellin, David Ho, Nancy Howell, Lorna E. Thorpe, Tara C. Smith, Marc N. Gourevitch, Linda Villarosa, Carl Zimmer, John Brownstein, Jim Kim, Samuel Scarpino, Jeremy Farrar, Andy Slavitt, Nancy Bristow, Anthony Fauci, Clive Thompson, Joon Yun, and my multi-talented cohost of the television series, David Olusoga. Max Roser and the team at Our World in Data provided invaluable help on all the "vital statistics" in this book—they are the true heirs to Graunt and Farr. Special thanks to my brother-in-law Manesh Patel for his expert advice—and to both my parents, for so much invaluable support in such a difficult year. Props to the SERJ group for their invaluable big-picture ruminations. And I'm grateful to Stewart Brand and Ryan Phelan for helping inspire the title.

A multi-platform project like this one depends on the contributions of so many different people and organizations, starting with the publishing team at Riverhead: my brilliant editor, Courtney Young, who managed to shepherd this book through a pregnancy and a pandemic; and my publisher, Geoffrey Kloske, who supported the project through all its many iterations. Thanks as well to Jacqueline Shost, Ashley Garland, May-Zhee Lim, and Ashley Sutton. I'm indebted to my longtime editor, Bill Wasik, and Jake Silverstein at the *New York Times Magazine* for seeing the potential of this project at a critical early stage. At PBS, Bill Gardner has been a tireless champion for getting these kinds of ideas on the screen since we worked together on *How We Got To Now*. My podcast producers Marshall Lewy and Nathalie Chicha helped me connect some of the historical stories here to the present-day realities of COVID-19. I'm grateful for the financial and editorial support from the foundations and individuals who helped us make the television series, especially Doron Weber at the Sloan Foundation, who not only provided a crucial backstop of encouragement at a low point in the process of making the show, but also helped me visualize a way of weaving today's crisis into the triumphs of the past. Thanks as well to the Arthur Vining Davis Foundation for their support of the series. My agents—Lydia Wills, Ryan McNeily, Sylvie Rabineau, Travis Dunlap, and Jay Mandel—took all the various swerves of this project in stride and somehow managed to steer it to a terrific destination in the end.

Finally, thanks to my wife and sons, who truly make all this wonderful extra life worth living.

NOTES

INTRODUCTION: TWENTY THOUSAND DAYS

1. Starmans.
2. Erkorcka, 190–94.
3. Barry, 176.
4. Quoted in Opdycke, 108.
5. Barry, 397.
6. "In South African cities, those between the ages of twenty and forty accounted for 60 percent of the deaths. In Chicago the deaths among those aged twenty to forty almost quintupled deaths of those aged forty-one to sixty. A Swiss physician 'saw no severe case in anyone over 50.'" Barry, 238.
7. Quoted in Barry, 364.
8. Data courtesy of *Our World in Data*. https://ourworldindata.org /grapher/life-expectancy.
9. Data courtesy of *Our World in Data*. https://ourworldindata.org /grapher/child-mortality-around-the-world.
10. Bernstein, Lenny. "U.S. Life Expectancy Declines Again, A Dismal Trend Not Seen Since World War I." *Washington Post*, November

29, 2018. www.washingtonpost.com/national/health-science/us-life-ex
pectancy-declines-again-a-dismal-trend-not-seen-since-world-war-i
/2018/11/28/ae58bc8c-f28c-11e8-bc79-68604ed88993_story
.html?utm_term=.382543252d3c.

11. Fogel, "Catching Up with the Economy," 2.

CHAPTER 1: THE LONG CEILING: MEASURING LIFE EXPECTANCY

1. Voyage details based on an interview with Nancy Howell.

2. Howell, *Life Histories of the Dobe !Kung*, 1–3.

3. Howell, *Demography of the Dobe !Kung*. loc. 535–38.

4. Sahlins.

5. Graunt, 41.

6. Graunt, 72.

7. Graunt, 135.

8. Howell, *Demography of the Dobe !Kung*, loc. 872–76.

9. Howell, *Demography of the Dobe !Kung*, loc. 851–55.

10. Howell, *Demography of the Dobe !Kung*, loc. 980–96.

11. Quoted in Devlin, 97.

12. Deaton, 81.

13. Hollingsworth, 54.

14. It is possible that some other sustained upward trend in life expectancy occurred at an earlier point in some other society, but was simply not measured because that society did not keep the necessary records to track it. But what we know about the history of medicine and public health suggests that this is unlikely. What we do know for certain is that whatever sustained increase might have happened in the past proved to be fleeting, and had disappeared by the time accurate mortality data began to be recorded in countries around the world.

15. Deaton, 163.

16. Quoted in Hadlow, 358.

17. Hadlow, 359.

18. Cox et al., 334.

19. Hollingsworth, 54.

20. Quoted in Rosen, 5–6.

21. Osler, 135.

22. McKeown, *The Role of Medicine*. x.

23. McKeown, *The Modern Rise of Population*. 15.

CHAPTER 2: THE CATALOGUE OF EVILS:
VARIOLATION AND VACCINES

1. Needham, 124–34.
2. Razzell, 115.
3. Quoted in Hopkins, 206.
4. According to Dr. Mead, the main objective of their interventions was "to keep the inflammation of the blood within due bounds, and at the same time to assist the expulsion of the morbific matter through the skin." Quoted in Carrell, 47.
5. Quoted in Carrell, 73.
6. Quoted in Carrell, 82.
7. Gross and Sepkowitz, 54.
8. Gross and Sepkowitz, 54.
9. Quoted in James, 25.
10. Quoted in Leavell, 122.
11. Jefferson to John Vaughn, November 5, 1801. https://founders.archives.gov/documents/Jefferson/01-35 02-0464.
12. Quoted in Leavell, 124.
13. "Government Regulation." *The History of Vaccines.* https://www.historyofvaccines.org/index.php/content/articles/government-regulation.
14. Dickens, Charles. *Household Words Almanac.* djo.org.uk/household-words-almanac/year-1857/page-19.html.
15. Wolfe, 430-32.
16. Quoted in Fee and Fox, 107.
17. Quoted in Henderson, 884.
18. Quoted in Leavell, 122.
19. Foege, 76.

CHAPTER 3: VITAL STATISTICS:
DATA AND EPIDEMIOLOGY

1. Luckin, 33.
2. Quoted in Eyler, *Victorian Social Medicine*, 43.
3. Eyler, 29.
4. Eyler, 82.
5. Eyler, 92–95.
6. See Johnson, *The Ghost Map.*
7. Quoted in Eyler, 156.
8. Lewis, 132.

9. Du Bois, 36.
10. Du Bois, 204–205.
11. Du Bois, 328.

CHAPTER 4: SAFE AS MILK:
PASTEURIZATION AND CHLORINATION

1. Leslie, 18.
2. Quoted in Smith-Howard, 16.
3. Nelson.
4. Szreter, 25–26.
5. Hartley, 133.
6. *Frank Leslie's Illustrated Newspaper*, May 22, 1858. https://upload
   .wikimedia.org/wikipedia/commons/f/f6/Frank_Leslie%27s
   _Illustrated_Newspaper%2C_May_22%2C_1858%2C_front_page.jpg.
7. Quoted in Moss.
8. Dillon, 23.
9. "Pure Milk for the Poor." *New York Times*, May 16, 1894. https://
   timesmachine.nytimes.com/timesmachine/1894/05/16/106905450
   .html?pageNumber=9.
10. Smith-Howard, 22.
11. Straus, 98.
12. Smith-Howard, 33.
13. For more on Leal's work, see Johnson, *How We Got to Now* (New York:
    Riverhead, 2014) and McGuire, *The Chlorine Revolution*.
14. "What's Behind NYC's Drastic Decrease in Infant Mortality Rates?"
15. Cutler and Miller, 2004, 13-15.
16. Straus, 116.
17. Ruxin, 395.
18. "Miracle Cure for an Old Scourge."
19. "Miracle Cure for an Old Scourge."
20. "Miracle Cure for an Old Scourge."
21. Gawande.
22. Ruxin, 396.

CHAPTER 5: BEYOND THE PLACEBO EFFECT:
DRUG REGULATION AND TESTING

1. Quoted in Rosen, 242.
2. Barry, Kindle edition, 23.

3. Ballentine, 3–4.
4. "'Death Drug Hunt' Covered 15 States." *New York Times*, November 26, 1937, 42. https://timesmachine.nytimes.com/timesmachine/1937/11/26/94467337.html?action=click&contentCollection=Archives&module=ArticleEndCTA&region=ArchiveBody&pgtype=article&pageNumber=42.
5. Quoted in West.
6. Kelsey, 13.
7. Kelsey, 59.
8. Fisher, 49.
9. Hill, 1952, 113–19.
10. Doll and Hill, 743.
11. Eldridge, Lynne. "What Percentage of Smokers Get Lung Cancer?" VerywellHealth, June 26, 2020. verywellhealth.com/what-percentage-of-smokers-get-lung-cancer-2248868.
12. All quotes from Richard Doll from this 2004 interview: https://cancerworld.net/senza-categoria/richard-doll-science-will-always-win-in-the-end.

CHAPTER 6: THE MOLD THAT
CHANGED THE WORLD: ANTIBIOTICS

1. Quoted in Rosen, 94–95.
2. Williams, 162–65.
3. Quoted in Bendiner, 283.
4. Rosen, 45.
5. Macfarlane, 203.
6. Rosen, 123–25.
7. Quoted in Lax, 186.
8. Quoted in Lax, 190.
9. Technically, Chain had overseen an earlier penicillin experiment involving a cancer patient, though it wasn't designed to cure the cancer.
10. Quoted in "Committee on the History of the New York Section of the American Chemical Society 2007 Annual Report."
11. Farris.
12. Wainwright, 190.
13. Wainwright, 193.

CHAPTER 7: EGG DROPS AND ROCKET SLEDS: AUTOMOBILE AND INDUSTRIAL SAFETY

1. "Mary Ward, the First Person to be Killed in a Car Accident—31 August 1869." blog, August 30, 2013. https://blog.britishnewspaperarchive .co.uk/2013/08/30/mary-ward-the-first-person-to-be-killed -in-a-car-accident-31-august-1869.
2. Laskow.
3. Gangloff, 40.
4. DeHaven (1942), 5.
5. DeHaven (1942), 8.
6. Stapp, 100.
7. "The Man Behind High-Speed Safety Standards."
8. Quoted in Ryan, 107.
9. Quoted in Nader, 60.
10. Borroz.
11. Nader, 10.
12. Quoted in the 1965 *Congressional Quarterly*, 783.
13. United States Congress. *Congressional Record*. October 21, 1966. Vol. 112, 28618. https://www.google.com/books/edition/Congressional _Record/FBEb4lvtxMMC?hl=en&gbpv=1&dq=%22crusading+spirit +of+one+individual+who+believed+he+could+do+something%22&pg =PA28618&printsec=frontcover.
14. Based on data compiled by the National Safety Council, available at: injuryfacts.nsc.org/motor-vehicle/historical-fatality-trends/deaths-and -rates.

CHAPTER 8: FEED THE WORLD: THE DECLINE OF FAMINE

1. See Kauffman, and Johnson, *Where Good Ideas Come From*.
2. Majd, 17.
3. Majd, 23.
4. Fogel, loc. 852.
5. McKeown, *The Modern Rise of Population*, 142.
6. McKeown, *The Modern Rise of Population*, 156.
7. Gráda, 10–24.
8. For an excellent overview of the data on the decline of famine, see https://ourworldindata.org/famines.
9. Adler, "How the Chicken Conquered the World."

10. Cushman, 40–43.
11. India Broiler Meat (Poultry) Production by Year. https://www
.indexmundi.com/agriculture/?country=in&commodity=broiler
-meat&graph=production.

CONCLUSION: BHOLA ISLAND, REVISITED

1. It is true that no large-scale modern society has approached the level of egalitarianism that Nancy Howell found in the !Kung societies. But it is a lot easier to be an egalitarian in a society that hasn't invented capital yet. There are only so many possessions you can maintain in a true hunter-gatherer culture. There were probably plenty of Paleolithic schemers who would have loved to turn into Steve Jobs (or Bernie Madoff, for that matter), but they couldn't because the adjacent possible of hunter-gatherer culture made it impossible to even imagine that kind of wealth accumulation. Many of the nations that have spent the most time tinkering with industrial capitalism—France, Netherlands, Germany—with all its promise and its peril, have seemingly settled on a model of democratic socialism, promarket but with a strong safety net and universal health care, with which it is possible to build successful national economies with impressive levels of equality. (The United States, alas, has not yet embraced that model.) There is good reason to suspect—based on the trends visible in the chart above—that those same outcomes are possible between nations as well, that the gradient in both wealth and longevity will continue to shrink over the coming decades.
2. Fogel, *The Escape from Hunger and Premature Death*, loc. 804–18.
3. City Health Dashboard. https://www.cityhealthdashboard.com.
4. D'Agostino.
5. See https://www.who.int/data/gho/data/themes/malaria.
6. Hammond et al., 80–83.
7. Richtel, 298–300.
8. Research Data Alliance. https://www.rd-alliance.org/groups/rda
-covid19.
9. HealthWeather. https://healthweather.us/
10. Johnson, "How Data Became."
11. Silver et al., 1140–42.
12. Somers.
13. Richtel, 248.

14. Adler et al., 3254–55.
15. "Turning Back Time."
16. Friend.
17. Cohen, 121.
18. Diamond, Jared. "The Worst Mistake in the History of the Human Race." *Discover*, May 1, 1999. http://discovermagazine.com/1987/may /02-the-worst-mistake-in-the-history-of-the-human-race.

BIBLIOGRAPHY

Adler, Adam S., et al. "Motif Module Map Reveals Enforcement of Aging by Continual NF-kB Activity." *Genes and Development* 21, no. 24 (2007), 3244–57, doi:10.1101/gad.1588507.

Adler, Jerry. "How the Chicken Conquered the World." *Smithsonian Magazine*, June 1, 2012, www.smithsonianmag.com/history/how-the-chicken-conquered-the-world-87583657/#IfRbIAss4zRjbFBE.99.

Aldrich, Mark. "History of Workplace Safety in the United States, 1880–1970." *EHnet.* www.eh.net/encyclopedia/history-of-workplace-safety-in-the-united-states-1880-1970.

Anderson, D. Mark, et al. "Public Health Efforts and the Decline in Urban Mortality: Reply to Cutler and Miller." *SSRN Electronic Journal*, 2019, doi:10.2139/ssrn.3314366.

Armitage, Peter. "Fisher, Bradford Hill, and Randomization." *International Journal of Epidemiology* 32, no. 6 (2003), 925–28, doi:10.1093/ije/dyg286.

———. "Obituary: Sir Austin Bradford Hill, 1897–1991." *Journal of the Royal Statistical Society: Series A (Statistics in Society)* 154, no. 3 (1991), 482–84, doi:10.1111/j.1467-985x.1991.tb00329.x.

Ballentine, Carol. "Taste of Raspberries, Taste of Death: The 1937 Elixir Sulfanilamide Incident." *FDA Consumer*, June 1981.

Barry, John M. *The Great Influenza: The Story of the Deadliest Pandemic in History*. New York: Penguin Books, 2018.

Bendiner, Elmer. "Alexander Fleming: Player with Microbes." *Hospital Practice* 24, no. 2 (1989), 283–316, doi:10.1080/21548331.1989.117 03671.

Bloom, David E., et al. "The Value of Vaccination," in *Fighting the Diseases of Poverty*, edited by Philip Stevens. New York: Routledge, 2017, 214–38.

Borroz, Tony. "Strapping Success: The 3-Point Seatbelt Turns 50." *Wired*, June 4, 2017. www.wired.com/2009/08/strapping-success-the-3-point -seatbelt-turns-50.

Boylston, Arthur. "The Origins of Inoculation." *Journal of the Royal Society of Medicine* 105, no. 7 (2012), 309–13, doi:10.1258/jrsm.2012 .12k044.

Bulletin of the World Health Organization. "Miracle Cure for an Old Scourge. An Interview with Dr. Dhiman Barua." March 4, 2011, www .who.int/bulletin/volumes/87/2/09-050209/en.

Burroughs Wellcome and Company. *The History of Inoculation and Vaccination for the Prevention and Treatment of Disease*. Lecture memoranda, Australasian Medical Congress, Auckland, N.Z., 1914.

Carrell, Jennifer Lee. *The Speckled Monster: A Historical Tale of Battling Smallpox*. New York: Plume, 2004.

Ciecka, James E. "The First Probability Based Calculations of Life Expectancies." *Journal of Legal Economics* 47 (2011), 47–58.

Cohen, Mark Nathan. *Health and the Rise of Civilization*. New Haven, CT: Yale University Press, 2011.

"Committee on the History of the New York Section of the American Chemical Society 2007 Annual Report." American Chemical Society. www.newyorkacs.org/reports/NYACSReport2007/NYHistory.html.

Cox, Timothy M., et al. "King George III and Porphyria: An Elemental Hypothesis and Investigation." *The Lancet* 366, no. 9482 (2005), 332–35, doi:10.1016/s0140-6736(05)66991-7.

Cushman, G. T. *Guano and the Opening of the Pacific World: A Global Ecological History*. Cambridge, UK: Cambridge University Press, 2013.

Cutler, David, and Grant Miller. "The Role of Public Health Improvements in Health Advances: The Twentieth-Century United States." *Demography* 42 (2005), 1-22, doi:10.3386/w10511.

Cutler, David, et al. "The Determinants of Mortality." *Journal of Economic Perspectives* 20, no. 3 (Summer 2006), 97-120, doi:10.3386/w11963.

D'Agostino, Ryan. "How Does Bill Gates's Ingenious, Waterless, Life-Saving Toilet Work?" *Popular Mechanics*, November 7, 2018. www .popularmechanics.com/science/health/a24747871/bill -gates-life-saving-toilet.

Deaton, Angus. *The Great Escape: Health, Wealth, and the Origins of Inequality.* Princeton, NJ: Princeton University Press, 2015.

DeHaven, Hugh. "Mechanical Analysis of Survival in Falls from Heights of Fifty to One Hundred and Fifty Feet." *Injury Prevention* 6, no. 1, (2000), doi:10.1136/ip.6.1.62-b.

Dillon, John J. *Seven Decades of Milk: A History of New York's Dairy Industry.* Ann Arbor, MI: University of Michigan Press, 1993.

Doll, Richard, and A. Bradford Hill. "Smoking and Carcinoma of the Lung." *The British Medical Journal* 2, no. 4682 (1950), 739–748, doi:10.1136/bmj.2.4682.739.

Doll, Richard, and A. Bradford Hill. "The Mortality of Doctors in Relation to Their Smoking Habits." *The British Medical Journal* 1, no. 4877 (1954), 1451–55, doi:10.1136/bmj.1.4877.1451.

Du Bois, W. E. B. *The Philadelphia Negro* (The Oxford W. E. B. Du Bois). Kindle edition. New York: Oxford University Press, 2014.

Erkoreka, Anton. "Origins of the Spanish Influenza Pandemic (1918–1920) and Its Relation to the First World War." *Journal of Molecular and Genetic Medicine* 3, no. 2 (2009), doi:10.4172/1747-0862.1000033.

Eyler, John M. "Constructing Vital Statistics: Thomas Rowe Edmonds and William Farr, 1835–1845." In A. Morabia, ed., *A History of Epidemiologic Methods and Concepts.* Basel, Switzerland: Birkhäuser, 2004, 149–57, doi:10.1007/978-3-0348-7603-2_4.

———. *Victorian Social Medicine: The Ideas and Methods of William Farr.* Baltimore: Johns Hopkins University Press, 1979.

Farris, Chris. "Moldy Mary . . . Or a Simple Messenger Girl?" *Peoria Magazine*, December 2019. www.peoriamagazines.com/pm/2019/dec /moldy-mary-or-simple-messenger-girl.

Fee, Elizabeth, and Daniel M. Fox. *AIDS: The Making of a Chronic Disease.* Oakland, CA: University of California Press, 1992.

Fisher, Ronald Aylmer. *The Design of Experiments*, 3rd ed. London: Oliver and Boyd, 1942.

Foege, William H. *House on Fire: The Fight to Eradicate Smallpox.* Oakland, CA: University of California Press, 2012.

Fogel, Robert W. "Catching Up with the Economy." *American Economic Review* 89, no. 1 (1999), 1–21, doi:10.1257/aer.89.1.1.

———. *The Escape from Hunger and Premature Death, 1700–2100.* New York: Cambridge University Press, 2003.

Frerichs, Ralph R. "Reverend Henry Whitehead." www.ph.ucla.edu/epi /snow/whitehead.html.

Friend, Tad. "Silicon Valley's Quest To Live Forever." *New Yorker*, March 27, 2017. www.newyorker.com/magazine/2017/04/03/silicon-valleys -quest-to-live-forever.

Galloway, James N., et al. "A Chronology of Human Understanding of the Nitrogen Cycle." *Philosophical Transactions of the Royal Society B: Biological Sciences* 368, no. 1621 (2013), 20130120, doi:10.1098/rstb.2013.0120.

Gammino, Victoria M. "Polio Eradication, Microplanning and GIS." *Directions Magazine*—GIS News and Geospatial, July 16, 2017. www .directionsmag.com/article/1350.

Gammino, Victoria M., et al. "Using Geographic Information Systems to Track Polio Vaccination Team Performance: Pilot Project Report." *Journal of Infectious Diseases* 210, issue suppl. 1 (2014), doi:10.1093/infdis/ jit285.

Gawande, Atul. "Slow Ideas." *New Yorker*, July 22, 2013. www.newyorker .com/magazine/2013/07/29/slow-ideas.

Gelfand, Henry M., and Joseph Posch. "The Recent Outbreak of Smallpox in Meschede, West Germany." *American Journal of Epidemiology* 93, no. 4 (1971), 234–37, doi:10.1093/oxfordjournals.aje.a121251.

Glass, D. V. "John Graunt and His Natural and Political Observations." *Notes and Records of the Royal Society of London* 19, no. 1 (1964), 63–100, doi:10.1098/rsnr.1964.0006.

Godfried, Isaac. "A Review of Recent Reinforcement Learning Applications to Healthcare." Medium, *Towards Data Science*, January 9, 2019.

Gráda, Cormac Ó. *Famine: A Short History.* Princeton, NJ: Princeton University Press, 2010.

Graunt, John. *Natural and Political Observations: Mentioned in a Following Index and Made upon the Bills of Mortality; With Reference to the Government, Religion, Trade, Growth, Air, Diseases, and the Several Changes of the Said City.* London: Martyn, 1676.

Griffith, G. Talbot. *Population Problems of the Age of Malthus.* Cambridge, UK: Cambridge University Press, 2010.

Gross, Cary P., and Kent A. Sepkowitz. "The Myth of the Medical Breakthrough: Smallpox, Vaccination, and Jenner Reconsidered." *International*

Journal of Infectious Diseases 3, no. 1 (1998), 54–60, doi:10.1016/s1201
-9712(98)90096-0.

Guerrant, Richard L., et al. "Cholera, Diarrhea, and Oral Rehydration
Therapy: Triumph and Indictment." *Clinical Infectious Diseases* 37, no.
3 (2003), 398–405, doi:10.1086/376619.

Habakkuk, H. J. *Population Growth and Economic Development since 1750.*
Leicester, UK: Leicester University Press, 1981.

Hadlow, Janice. *A Royal Experiment: The Private Life of King George III.*
New York: Henry Holt and Company, 2014.

Hammond, Andrew, et al. "A CRISPR-Cas9 Gene Drive System Targeting
Female Reproduction in the Malaria Mosquito Vector *Anopheles Gam-
biae.*" *Nature Biotechnology* 34, no. 1 (2016), 78–83, doi:10.1038/nbt
.3439.

Handley, J. B. "The Impact of Vaccines on Mortality Decline Since 1900—
According to Published Science," Children's Health Defense, March 12,
2019. www.childrenshealthdefense.org/news/the-impact-of-vaccines-on
-mortality-decline-since-1900-according-to-published-science.

Harris, Bernard. "Public Health, Nutrition, and the Decline of Mortality:
The McKeown Thesis Revisited." *Social History of Medicine* 17, no. 3
(2004) 379–407.

Hartley, Robert Milham. *An Historical, Scientific, and Practical Essay on
Milk, as an Article of Human Sustenance: With a Consideration of the
Effects Consequent upon the Present Unnatural Methods of Producing It
for the Supply of Large Cities.* London: Forgotten Books, 2016.

Henderson, Donald A. "A History of Eradication: Successes, Failures, and
Controversies." *The Lancet* 379, no. 9819 (2012), 884–5.

Hill, A. Bradford. "The Clinical Trial." *New England Journal of Medicine*
247, no. 4 (1952), 113–19.

Hollingsworth, T. H. "Mortality." *Population Studies* 18, no. 2 (November
1964).

Hopkins, Donald R. *The Greatest Killer: Smallpox in History*, with a new
introduction. Chicago: University of Chicago Press, 2002.

Howell, Nancy. *Demography of the Dobe !Kung.* Kindle edition. New York:
Routledge, 2007.

———. *Life Histories of the Dobe !Kung: Food, Fatness, and Well-Being
over the Life Span.* Oakland, CA: University of California Press,
2010.

Hull, Charles H. "Graunt or Petty?" *Political Science Quarterly* 11, no. 1
(1896), 105, doi:10.2307/2139604.

James, Portia P. *The Real McCoy: African-American Invention and Innovation, 1619–1930.* Washington, DC: Smithsonian Institution Press, 1990.

Jha, Prabhat, and Witold A. Zatonski, "Smoking and Premature Mortality: Reflections on the Contributions of Sir Richard Doll." *Canadian Medical Association Journal* 173, no. 5 (2005), 476–77, doi:10.1503/cmaj.050948.

Johnson, Steven. "How Data Became One of the Most Powerful Tools to Fight an Epidemic." *New York Times Magazine*, June 11, 2020. www.nytimes.com/interactive/2020/06/10/magazine/covid-data.html.

———. *The Ghost Map: The Story of London's Most Terrifying Epidemic— and How It Changed Science, Cities, and the Modern World.* New York: Riverhead, 2006.

———. *Where Good Ideas Come From: The Natural History of Innovation.* New York: Riverhead, 2011.

Kauffman, Stuart A. *Investigations.* New York: Oxford University Press, 2002.

Laskow, Sarah. "Railyards Were Once So Dangerous They Needed Their Own Railway Surgeons." *Atlas Obscura*, July 25 2018. www.atlasobscura.com/articles/what-did-railway-surgeons-do.

Lax, Eric. *The Mold in Dr. Florey's Coat: The Story of the Penicillin Miracle.* New York: Henry Holt, 2005.

Leavell, B. S. "Thomas Jefferson and Smallpox Vaccination." *Transactions of the American Clinical and Climatological Association* 88 (1977), 119–27.

Leslie, Frank. *The Vegetarian Messenger* 10 (1858).

Lewis, David L. *W.E.B. Du Bois: A Biography, 1868–1963.* Kindle edition. New York: Henry Holt and Company, 2009.

Luckin, W. "The Final Catastrophe—Cholera in London, 1866." *Medical History* 21, no. 1 (1977) 32–42, doi:10.1017/s0025727300037157.

Macfarlane, Gwyn. *Howard Florey: The Making of a Great Scientist.* The Scientific Book Club, 1980.

Majd, Mohammad Gholi. *The Great Famine and Genocide in Persia, 1917– 1919.* Lanham, MD: University Press of America, 2003.

"The Man Behind High-Speed Safety Standards." National Air and Space Museum, August 22, 2018. www.airandspace.si.edu/stories/editorial/man-behind-high-speed-safety-standards.

McGuire, Michael J. *The Chlorine Revolution: The History of Water Disinfection and the Fight to Save Lives.* American Denver: Water Works Association, 2013.

McKeown, Thomas. *The Role of Medicine: Dream, Mirage, or Nemesis?* Princeton, NJ: Princeton University Press, 2016.

———. *The Modern Rise of Population*. London: Edward Arnold, 1976.

McNeill, Leila. "The Woman Who Stood Between America and a Generation of 'Thalidomide Babies'." *Smithsonian Magazine*, May 8, 2017. www.smithsonianmag.com/science-nature/woman-who-stood-between-america-and-epidemic-birth-defects-180963165.

Moss, Tyler. "The 19th-Century Swill Milk Scandal That Poisoned Infants with Whiskey Runoff." *Atlas Obscura*, November 27, 2017. www.atlasobscura.com/articles/swill-milk-scandal-new-york-city.

Nader, Ralph. *Unsafe at Any Speed: The Designed-In Dangers of the American Automobile*. New York: Knightsbridge Publishing Co., 1991.

Najera, Rene F. "Black History Month: Onesimus Spreads Wisdom That Saves Lives of Bostonians During a Smallpox Epidemic." History of Vaccines. historyofvaccines.org/content/blog/onesimus-smallpox-boston-cotton-mather.

Needham, Joseph. "Biology and Biological Technology." *Science and Civilisation in China* 6, Part VI, Medicine. Cambridge, UK: Cambridge University Press, 2000.

Nelson, Bryn. "The Lingering Heat over Pasteurized Milk." Science History Institute, April 18, 2019. www.sciencehistory.org/distillations/the-lingering-heat-over-pasteurized-milk.

Opdycke, Sandra. *The Flu Epidemic of 1918: America's Experience in the Global Health Crisis*. New York: Routledge, 2014.

Osler, William. *The Principles and Practice of Medicine*, 8th ed., Largely Rewritten and Thoroughly Revised with the Assistance of Thomas McCrae. Boston: D. Appleton & Company, 1912.

Parke, Davis & Company. *1907–8 Catalogue of the Products of the Laboratories of Parke, Davis & Company, Manufacturing Chemists, London, England*. wellcomecollection.org/works/w5g9s5ac.

Pinker, Steven. *Enlightenment Now: The Case for Reason, Science, Humanism, and Progress*. New York: Penguin Books, 2019.

Plough, Alonzo L. *Advancing Health and Well-Being: Using Evidence and Collaboration to Achieve Health Equity*. New York: Oxford University Press, 2018.

"Policy Impact: Seat Belts." Centers for Disease Control and Prevention, January 3, 2011.

Pordeli, Mohammad Reza, et al. A Study of the Causes of Famine in Iran during World War I." *Review of European Studies* vol. 9, no. 2 (2017), p. 296, doi:10.5539/res.v9n2p296.

Razzell, Peter Ernest. *The Conquest of Smallpox: The Impact of Inoculation on Smallpox Mortality in Eighteenth Century Britain.* London: Caliban Books, 2003.

Richtel, Matt. *An Elegant Defense: The Extraordinary New Science of the Immune System: A Tale in Four Lives.* New York: William Morrow, 2020.

Riedel, Stefan. "Edward Jenner and the History of Smallpox and Vaccination." *Baylor University Medical Center Proceedings* 18, no. 1, (2005), 21–25, doi:10.1080/08998280.2005.11928028.

Riley, James C. *Rising Life Expectancy: A Global History.* New York: Cambridge University Press, 2015.

Rosen, William. *Miracle Cure: The Creation of Antibiotics and the Birth of Modern Medicine.* New York: Penguin Books, 2018.

Ruxin, Joshua Nalibow. "Magic Bullet: The History of Oral Rehydration Therapy." *Medical History* 38, no. 4 (1994), 363–97, doi:10.1017/s0025727300036905.

Ryan, Craig. *Sonic Wind: The Story of John Paul Stapp and How a Renegade Doctor Became the Fastest Man on Earth.* New York: Liveright, 2016.

Sahlins, Marshall. "The Original Affluent Society." Eco Action, 2005. www.eco-action.org/dt/affluent.html.

Saul, Toby. "Inside the Swift, Deadly History of the Spanish Flu Pandemic." *National Geographic*, March 5, 2020. www.nationalgeographic.com/history/magazine/2018/03-04/history-spanish-flu-pandemic.

Schultz, Stanley G. "From a Pump Handle to Oral Rehydration Therapy: A Model of Translational Research." *Advances in Physiology Education* 31, no. 4 (2007), 288–93, doi:10.1152/advan.00068.2007.

Silver, David, et al. "A General Reinforcement Learning Algorithm That Masters Chess, Shogi, and Go Through Self-Play." *Science* 362, no. 6419 (2018), 1140–44, doi:10.1126/science.aar6404.

Smith-Howard, Kendra. *Pure and Modern Milk: An Environmental History since 1900.* New York: Oxford University Press, 2017.

Somers, James. "How the Artificial Intelligence Program AlphaZero Mastered Its Games." *New Yorker*, December 28, 2018. www.newyorker

.com/science/elements/how-the-artificial-intelligence-program
-alphazero-mastered-its-games.

Stapp, J. P. "Problems of Human Engineering in Regard to Sudden Decelerative Forces on Man." *Military Medicine* 103, no. 2 (1948), 99–102, doi:10.1093/milmed/103.2.99.

Starmans, Barbara J. "Spanish Influenza of 1918." *The Social Historian*, September 7, 2015. www.thesocialhistorian.com/spanish-influenza-of -1918.

Straus, Nathan. *Disease in Milk: The Remedy Pasteurization: The Life Work of Nathan Straus.* Smithtown, NY: Straus Historical Society, Inc., 2016.

Szreter, Simon. "The Importance of Social Intervention in Britain's Mortality Decline c.1850–1914: A Re-Interpretation of the Role of Public Health." *Social History of Medicine* 1, no. 1 (1988), 1–38, doi:10.1093/ shm/1.1.1

"Turning Back Time: Salk Scientists Reverse Signs of Aging." *Salk News*, December 15, 2016. www.salk.edu/news-release/turning-back-time-salk -scientists-reverse signs-aging.

"The Value of Vaccination." *The Lancet* 200, no. 5178 (1922), 1139, doi:10.1016/s0140-6736(01)01172-2.

Wagstaff, Anna. "Richard Doll: Science Will Always Win in the End." *Cancerworld*, November 23, 2017. www.cancerworld.net/senza-categoria /richard-doll-science-will-always-win-in-the-end/.

Wainwright, Milton. "Hitler's Penicillin." *Perspectives in Biology and Medicine* 47, no. 2 (2004), 189–198, doi:10.1353/pbm.2004.0037

West, Julian G. "The Accidental Poison That Founded the Modern FDA." *The Atlantic*, January 16, 2018. www.theatlantic.com/technology/-2018 /01/the-accidental-poison-that-founded-the-modern-fda/550574.

"What's Behind NYC's Drastic Decrease in Infant Mortality Rates?" National Institute for Children's Health Quality, July 24, 2017. www .nichq.org/insight/whats-behind-nycs-drastic-decrease-infant-mortality-rates.

Whitehead, M. "William Farr's Legacy to the Study of Inequalities in Health." *Bulletin of the World Health Organization*, 2000. www.ncbi .nlm.nih.gov/pmc/articles/PMC2560600.

Williams, Max. *Reinhard Heydrich: The Biography, Volume 2: Enigma.* Church Stretton, UK: Ulric Publishing, 2003.

Winter, R. et al. "Deep Learning for De Novo Drug Design." Interdisziplinärer Kongress | Ultraschall 2019–43. Dreiländertreffen DEGUM | ÖGUM | SGUM, 2019, doi:10.1055/s-0039-1695913.

Wolfe, Robert M., and Lisa Sharp. "Anti-Vaccinationists Past and Present." *BMJ* 325, no. 7361 (2002), 430–32, doi:10.1136/bmj.325.7361.430.

Zaimeche, Salah, and Salim Al-Hassani. "Lady Montagu and the Introduction of Smallpox Inoculation to England." *Muslim Heritage*, February 16, 2010. muslimheritage.com/lady-montagu-smallpox-inoculation-england.

INDEX